NIGHT SKY GUIDE 2026

North America's Skies Month by Month from
The Royal Astronomical Society of Canada

Nicole Mortillaro

FIREFLY BOOKS

A Firefly Book

Published by Firefly Books Ltd. 2025
Copyright © 2025 Firefly Books Ltd.
Copyright © 2025 The Royal Astronomical Society of Canada
Text © 2025 Nicole Mortillaro
Photographs © as listed on page 128

All rights reserved. No part of this publication may be reproduced, stored in a retrieval system, or transmitted in any form or by any means, electronic, mechanical, photocopying, recording or otherwise, without the prior written permission of the Publisher.

First printing

ISBN: 978-0-2281-0586-2

Published in Canada by
Firefly Books Ltd.
50 Staples Avenue, Unit 1
Richmond Hill, Ontario L4B 0A7

Published in the United States by
Firefly Books (U.S.) Inc.
P.O. Box 1338, Ellicott Station
Buffalo, New York 14205

Project manager/Editor (Firefly Books): Julie Takasaki
Project manager (RASC): Robyn Foret, Past President of The RASC
Technical editors: James S. Edgar FRASC (Editor, RASC *Observer's Handbook*)
    and Chris Vaughan
Graphic design: Noor Majeed and Stacey Cho
Illustrations and sky charts: Peter Kovalik

Printed in China | E

We gratefully acknowledge the financial support of the Government of Canada for our publishing program.

Contents

Introduction	4
Handy Sky Measures	6
The Night Sky: A Cosmic Time Machine	8
Binoculars and Telescopes	10
Stars	14
Constellations	16
Comets, Asteroids and Meteors	18
Meteor Showers in 2026	22
The Moon	24
Observing the Moon	25
The Sun	28
Observing the Sun	30
Eclipses	31
Eclipses in 2026	32
The Northern Lights	36
The Planets	38
Deep-Sky Objects	42
Galaxies	44
The Sky Month-by-Month	
Introduction	46
January	48
February	54
March	60
April	66
May	72
June	78
July	84
August	90
September	96
October	102
November	108
December	114
The Messier Catalogue	121
Glossary	126
Resources	128
Photo Credits	128

Introduction

If you're reading this, you clearly love the night sky and all the joy it brings you.

This guide aims to provide novice and intermediate amateur astronomers with the knowledge they need to enjoy the night sky in 2026, from understanding planets, nebulae (clouds of gas and dust), comets, asteroids and meteors, to learning about annual and significant celestial events. This is your pocket guide to the cosmos.

But first, it's important that you understand how we navigate the night sky. Just like for navigation on Earth, astronomers use particular coordinates in the sky to figure out what we're looking at.

The **celestial sphere** is an imaginary sphere with Earth at its center. At any one time, an observer's night-sky view only includes half of this sphere, because the other half is below the horizon.

Earth's axis is tilted at 23.5 degrees to the **plane** of the Solar System — the plane being the orbit of the Earth around the Sun. For observers on the ground, the celestial sphere seems to rotate from east to west; that is why the Sun, for example, rises in the east and sets in the west.

The Earth's axis points less than 1 degree away from **Polaris**, the North Star, in the Northern Hemisphere. It's really important to know where north is, so you can navigate around the sky. Use a star chart, your smartphone or both to find true north at your observing location. (In the south, the Earth's axis points about one

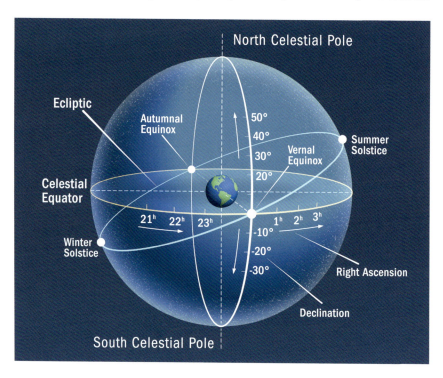

4 Introduction

degree away from **Sigma Octantis**, a faint star; given Canada and the United States are in the Northern Hemisphere, however, we will focus on Polaris.) When you look at the sky over long periods of time, Polaris stays still as the other stars rotate around it.

You can easily see this phenomenon by setting up a camera, pointing it toward Polaris, and taking an extended (or long-exposure) photograph of the stars. You will see circular streaks in the sky as other stars rotate around Polaris. Some stars are so close to the North Celestial Pole that they never set. Other stars farther from the pole will rise in the east and set in the west, just like our Sun. Many stars are so far from the pole that they never rise above your horizon.

While beginner astronomers will navigate using a simplified star chart, it's helpful to know some of the terminology we use for finding our way around the sky. That way, as you gain confidence, you can follow professional astronomers in their observations.

The point directly overhead your observing location is called the **zenith**, and the **celestial equator** is directly above the Earth's equator. The point directly below you (under the horizon and opposite to the zenith) is the **nadir**. The line that runs from the north point on the horizon, through the zenith, to the south point on the horizon, is the **meridian**.

To navigate the night sky, astronomers created a type of latitude and longitude called **celestial coordinates**. In astronomy, we use **declination** (Dec.) and **right ascension** (RA). Declination is like latitude on Earth, running from north to south. Right ascension is like longitude on Earth, running from west to east. Declination is measured in degrees, and right ascension is measured in hours, minutes and seconds.

Star trails with Polaris at the center

As Earth rotates, the apparent position of most of the stars changes. More advanced astronomers may want to learn more about exactly where to find faint stars in their telescopes. For these situations, we navigate using **altitude** (angular elevation above the horizon, between 0 degrees at the horizon and 90 degrees at the zenith) and **azimuth** (the number of degrees clockwise from due north). But if you're just starting out, don't worry yet about these advanced observing techniques. A simple star chart will help you find the constellations and the planets.

If you want to look for planets, you should also pay attention to the **ecliptic**. That is the path the Sun takes through the constellations, and the Moon and planets do not stray far from that path.

Introduction   **5**

Handy Sky Measures

Astronomers need to know how far apart things are in the sky. And they do this using angular degrees.

It's not hard to imagine the sky as a sphere that measures 360 degrees — after all, space surrounds Earth on all sides. Standing in one spot on Earth, if you trace the sky from horizon to horizon, that would equal 180 degrees. Remember, the other 180 degrees is under the horizon.

If you want to measure the distances between two objects — say, between the Moon and Venus as they appear together in the sky — you can use your hand as a measuring tool. It all lies within your fingers.

Hold your hand at arm's length. The width of your pinky finger equals 1 degree. The width of your three middle fingers held at arm's length equals five degrees; a closed fist is 10 degrees; the distance between the tip of your index finger and the tip of your pinky is 15 degrees; and the distance between your thumb and pinky is roughly 25 degrees.

This is particularly helpful when trying to see how high or low something is above the horizon.

You can practice measuring the degrees with stars found in the Big Dipper. The chart shows how to find the Big Dipper using Polaris. Because it's circumpolar, the orientation of the Big Dipper varies.

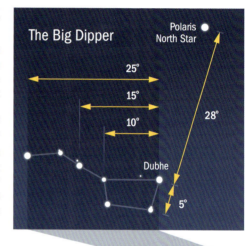

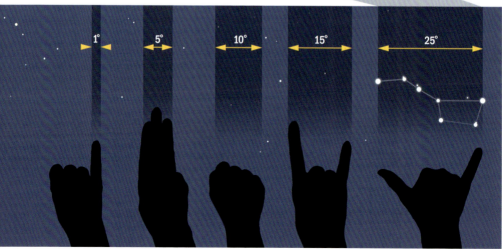

A view of Polaris and the Big Dipper over Dinosaur Provincial Park, Alberta

The Night Sky: A Cosmic Time Machine

Every time you look up at the sky, you are looking back in time.

Light from the Sun takes 8.3 minutes to reach us. Light from the Moon takes roughly 1.3 seconds. That's because light takes time to travel. In fact, light travels through space at roughly 300,000 kilometers (186,000 miles) per second.

We call the distance light travels in one year a **light-year**. So, when we look at stars, galaxies and nebulae, they appear how they looked back in time, depending on their distance to us. For example, the Andromeda Galaxy is the closest spiral galaxy to our own galaxy, the **Milky Way**. It is easily visible to the unaided eye in the Northern Hemisphere at dark-sky sites, even though it lies roughly 2.5 million light-years from us. That means, we are seeing the galaxy as it was 2.5 million years ago.

Sirius, also known as the "Dog Star," is the night sky's brightest star. It's visible during the winter months in the Northern Hemisphere. Its distance of 8.6 light-years from Earth means we are looking at it as it was 8.6 years ago.

Our Universe is roughly 13.8 billion years old. Astronomers improve their understanding of the evolution of our Universe by using powerful telescopes that can see farther and farther away, and so farther back in time.

Launched in 1990, the Hubble Space Telescope has provided us not only with

Andromeda Galaxy (Messier 31), our closest galactic neighbor

Captured by NASA's James Webb Space Telescope and titled "Webb's First Deep Field," at the time this image was the deepest and sharpest infrared image of the distant Universe. The curved arcs are galaxies distorted by gravitational lensing

jaw-dropping images of astronomical objects such as galaxies and nebulae, but it has also helped narrow down the age of the Universe.

And more recently, in 2021, the James Webb Space Telescope opened its eyes for the first time, providing us with a look back to some of the very first galaxies to form — back to when the Universe was less than a billion years old. It's almost like looking at a baby picture of our Universe.

While time machines may not be possible, every time we pick up binoculars or peer through our telescopes, we are taking a voyage back in time.

Binoculars and Telescopes

Using your eyes to move around the sky allows you to see the Moon, the planets, the Sun and even some objects outside of our Solar System. However, having binoculars or a telescope opens up the Universe to you.

Binoculars are an amazing first tool for seeking out the marvelous wonders of the night sky. Good binoculars can reveal the intricacies of the Moon, showing its peaks and valleys, or uncover the daily motion of Jupiter's moons as they dance around our Solar System's largest planet. A decent telescope can make these objects appear bigger and show details of star clusters or nebulae.

But the question that often arises is what type of binoculars or telescope should I get?

For binoculars, it's important to understand two things: the **magnification** (or power) and the **aperture**. Binoculars are usually represented by two numbers, separated by an "×." Two example binocular numbers are 7×50 or 10×50. The magnification is the first number, and it represents the number of times larger something will appear compared to viewing it with the naked eye. The second number is the aperture in millimeters (about 3/64ths of an inch), and it represents the diameter of each lens. Good viewing in part depends on how well your binoculars can gather light. If your binoculars have a larger second number, they have a larger aperture and can gather more light from distant objects. But that comes with a cost. Bigger binoculars are heavier and, therefore, more difficult to hold, which means you will need a tripod to steady your view.

Another number often given for binoculars is the **field of view**, or FOV. This is how wide you will be able to see in degrees (see "Handy Sky Measures," on page 6). In general, the higher the magnification you use, the smaller your field of view will be. At times, this will mean you need to make a choice. Do you want to zoom in on a particular crater on the Moon, for example, or observe a more spread-out chain of mountains? Making these decisions is difficult for astronomers, too, so don't worry if it feels hard — it gets

Binoculars are a great portable tool for observing the night sky

a bit easier to decide what to do with practice.

The top end of handheld binoculars is typically 7×50. Larger than that, it's best to get a tall tripod if you want the best view possible and to share views with others.

We recommend using binoculars for a few months before investing in a telescope. Once you are comfortable with binoculars, luckily, your knowledge will be useful because telescopes use many of the same definitions for observing.

Remember that light-gathering ability is important; a typical beginner's telescope usually ranges between 50 to 150 millimeters (2 to 6 inches) in **aperture**, the diameter of the main lens or mirror. The bigger the aperture, the more light the lens will gather, and the brighter the view will be. Invest in a sturdy tripod, too, so that the telescope will not shake when you use it.

The **magnification** of a telescope depends on the eyepiece used; magnification is calculated by dividing the focal length of the telescope by the focal length of the eyepiece (using like units). For example, a telescope with a focal length of 1,000 millimeters and a 10-millimeter eyepiece will magnify an object 100 times. The same 10-millimeter eyepiece in a 500-millimeter telescope would magnify an object 50 times.

While a bigger aperture and higher magnification may seem like the best way to go, it's best not to get too caught up in choosing a telescope with the highest numbers. What is important is how you plan to use it. Since most people are unlikely to have backyard observatories, the most important thing may be portability. Too heavy a telescope means you're unlikely to haul it out to the backyard or to a vacation spot, far from city lights.

When it comes to telescopes, there is a wide array of choices. Some of the most popular

A refractor telescope, which is best used for viewing planets, the Moon and double stars

are refractors, reflectors and compound telescopes. Each type has its own benefits, and it all depends on what you prefer. We therefore recommend you try testing out different telescopes at a local star party held by astronomy groups before making the investment. Alternatively, check reputable astronomy magazines or forums for recommendations for beginners; in most cases, all it takes is a little Internet searching and some patience.

A **refractor telescope** has a front lens that focuses light to form an image at the back, and an eyepiece that acts like a magnifying glass to allow your eye to focus on that image. Refractors tend to be more reliable, as their lenses are fixed in place and, therefore, don't get out of alignment as easily as some other types of telescope do. Refractors are best used for planetary and lunar observing as well as viewing double stars.

A **reflector telescope** uses a mirror to gather and focus light. The advantage for reflectors is they don't usually suffer from **chromatic aberrations**, when light of different wavelengths (i.e., colors) doesn't focus on the same point; this makes them ideal to observe distant objects like star clusters. Chromatic aberrations cause the different colors of the spectrum to split and the image to appear blurry, which isn't great when looking at a group of stars. A **Dobsonian telescope**, which is a variant of a reflector with a simple mount, gives you a far bigger aperture for far less than the cost of other telescopes.

Compound or **catadioptric telescopes** combine lenses and mirrors to form an image. The **Schmidt-Cassegrain**, a type of compound telescope, has a compact design that makes it quite popular. With this instrument, an astronomer can get a bigger aperture in a smaller sized telescope. This telescope type is also portable, making it easier to move the equipment into remote areas.

Star parties are great opportunities to check out different telescopes and chat with fellow enthusiasts

A reflector telescope (front) and a variant of a reflector telescope, a Dobsonian telescope (back), are ideal for viewing distant objects since they don't usually suffer from chromatic aberrations

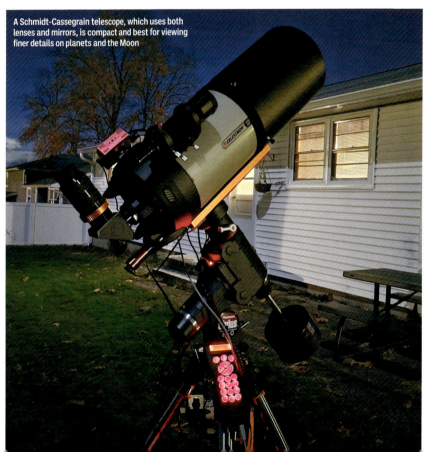

A Schmidt-Cassegrain telescope, which uses both lenses and mirrors, is compact and best for viewing finer details on planets and the Moon

Stars

Stars come in many different varieties. Our Sun is a **yellow dwarf star**, on the **main sequence**, meaning that it's converting hydrogen into helium at its core, like most other stars. When a star does this conversion, it releases a tremendous amount of energy. This energy, in the form of sunlight, allows life to thrive on Earth.

The Sun, at 4.5 billion years old, is considered middle-aged. When it runs out of fusion fuel, five billion years from now, it will at first swell, becoming a **red giant** and engulfing the inner planets, then it will slough off its outer layers to become a **white dwarf**.

One of the most interesting types of stars, some might argue, are **red supergiants**. These colossal stars — roughly 1,400 times the mass of the Sun — have relatively short lifespans.

And when supergiants do stop fusion, they do so in a spectacular fashion, in an explosion called a **supernova**. A supernova occurs when a star can no longer convert hydrogen into helium; eventually the core is converted into iron. During a supernova, the star first collapses and then explodes outward, creating even heavier elements.

Betelgeuse, a star found in the left shoulder of Orion, is a red supergiant. While astronomers regularly see supernovae occurring in far-away galaxies, it's been more than 400 years since one was observed within our own Milky Way galaxy. When Betelgeuse goes supernova, its brightness will rival the full Moon in our sky. If you're hoping to see Betelgeuse explode, you're not alone. However, estimates peg Betelgeuse's death for some time within the next 100,000 years.

The most common type of star in the Universe is a **red dwarf**, which is a cool star that's much smaller than the Sun. Astronomers often search for potentially habitable planets around these stars, because the inherent dimness and smaller size of red dwarfs makes it easier to spot planets. The stars, however, can be quite volatile, occasionally releasing a tremendous amount of radiation. Intense radiation is not a friendly process for most life forms.

The **Hertzsprung-Russell diagram** (H-R diagram) was developed in the early 1900s by Ejnar Hertzsprung and Henry Norris Russell, following the research findings by two Harvard computers, Annie Jump Cannon and Antonia Maury. The diagram plots the temperature of stars against their luminosity. Just as we do, stars go through

Top 10 Brightest Stars in the Night Sky	
1.	Sirius
2.	Canopus
3.	Arcturus
4.	Alpha Centauri A (Rigil Kentaurus)
5.	Vega
6.	Capella
7.	Rigel
8.	Procyon
9.	Achernar
10.	Betelgeuse

The Harvard Observatory Computers

In 1881, Edward Charles Pickering, the director of the Harvard Observatory, hired a team of women to compute and catalog photographs of the night sky. Their work was incredibly important in providing the foundations of astronomical theory. Annie Jump Cannon, one of the computers, added to work done by fellow computer Antonia Maury and developed a system of classifying stars that is still used today.

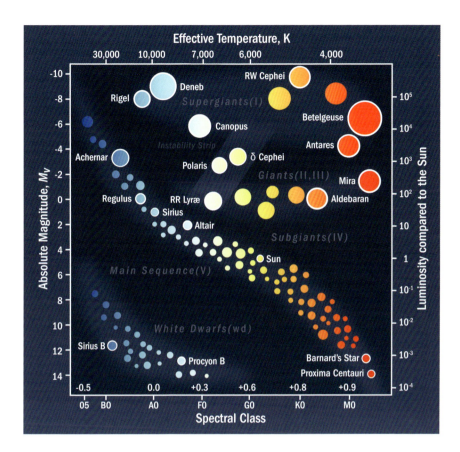

certain stages in their lives. The H-R diagram provides astronomers with the information about a star's current age. Main-sequence stars that are fusing hydrogen into helium — such as our Sun — lie on the diagonal branch of the diagram.

Finally, there's a star's **magnitude**, or apparent brightness. Magnitude is measured on a scale where the higher the number, the fainter it is — and negative numbers are brighter than positive numbers. For example, Sirius, the brightest star in the night sky, measures -1.4 on this scale. Polaris is 2, and the Sun is -27. We also use magnitude to measure the brightness of other celestial objects, like the Moon, planets, asteroids and comets.

Note there is a difference between apparent and absolute magnitude. Apparent magnitude is the brightness of an object that we observe from Earth, but absolute magnitude is the brightness of an object if it were placed 32.6 light-years from Earth. This second measure helps astronomers directly compare the luminosity of objects and is what is used for the H-R diagram. On this scale, Sirius has a magnitude of 1.4, Polaris is -3.6 and the Sun is 4.8.

In astronomy, Greek letters of the alphabet are used to identify stars within a constellation usually from brighter to dimmer.

Constellations

Constellations, groups of stars that make imaginary images in the night sky, have been around since ancient times. Typically, the images astronomers use are based on Greek, Roman and Arabic mythologies. The International Astronomical Union (IAU) recognizes a total of 88 official constellations.

Some of the most recognizable constellations in the Northern Hemisphere are Orion (the Hunter), Cygnus (the Swan), Leo (the Lion), Gemini (the Twins), Scorpius (the Scorpion) and Ursa Major (the Great Bear).

More recently, there has been more effort to acknowledge constellations of Indigenous peoples. The naming of Indigenous constellations varies around the world, making these constellations regionally distinct. Some groups see Ursa Major as a bear, or a caribou. The Cree see Corona Borealis as seven birds and Cepheus as a turtle. To the Navajo, Polaris is Nahookos Bikq, meaning central fire. For some Indigenous peoples, such as the Inuit, the Northern Lights are dancing spirits.

Aside from the constellations, there are also **asterisms**, a group of stars within a constellation (or sometimes from several different constellations) that forms its own distinct pattern. The Big Dipper is probably the most famous asterism, as its stars lie within the constellation of Ursa Major. There's also the Summer Triangle, with bright stars from Cygnus, Lyra (the Harp) and Aquila (the Eagle), and the Winter Triangle with stars from Orion, Canis Major (the Great Dog) and Canis Minor (the Little Dog).

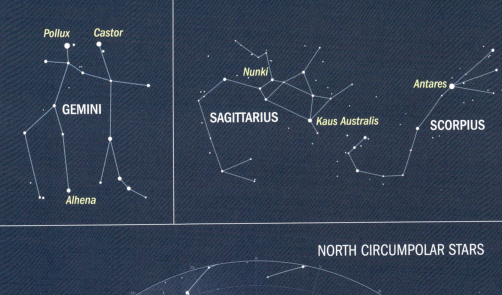

Comets, Asteroids and Meteors

Comets are icy balls of debris — specifically, dust and ice — left over from the formation of our Solar System. They are sometimes referred to as "dirty snowballs" and can be stunning objects to see in the night sky when tails of dust and ionized gas fan out behind their cores. While we know of many comets, predicting the appearance of a bright one is all but impossible. Comets tend to fall apart before becoming too bright. The Sun's gravity and heat are strong, and comets themselves are very delicate visitors from the outer Solar System or beyond. Many comets literally crumble under the pressure as they dive in toward the Sun and the inner Solar System, where our planet resides.

Bright comets can be marvelous. On August 17, 2014, Comet Lovejoy C/2014 Q2 was spotted as it came toward the inner Solar System. This comet produced a spectacular tail as it neared the Sun. Tails occur when ice **sublimates**, turning directly from a solid into a gas.

There have been other wonderful naked-eye comets that have graced our night sky, including Comet Hyakutake C/1996 B2, Hale-Bopp C/1995 O1, and Comet NEOWISE C/2020 F3. Each comet has the year of its discovery in its official name, so NEOWISE was found in 2020, Hyakutake in 1996, and so on. Many Northern Hemisphere observers were treated to a special show when Comet NEOWISE appeared in the skies in 2020. It was one of the brightest comets to appear in a generation — even city dwellers could spot it through light pollution.

Periodic comets, or ones that we can predict, are given the designation "P," while those that appear unexpectedly are given the designation "C." For example, Halley's Comet, or 1P/Halley, appears roughly every 76 years — a clear P. The last time it passed was 1986; the next time will be in 2061. In general, C comets are brighter than P comets because P comets have sublimated their material into space from repeated trips by the Sun.

Like comets, **asteroids** (which are called minor planets) are left over from the formation of our Solar System. Instead of dust and ice,

Comet NEOWISE C/2020 F3

asteroids are rocks and come in all shapes and sizes. Most travel in the asteroid belt between Mars and Jupiter; however, they can be found other places, too, including beyond the orbit of dwarf planet Pluto at the outer edge of the Solar System, in a region called the Kuiper Belt. There are three broad composition types of asteroids:

- **Chondrite (C-type)**, which are made up of clay and silicate rocks. They are dark in appearance and the most common. They are also the oldest type of asteroid.
- **Nickel-iron (M-type)**, which are believed to have experienced high temperatures and melted after they formed.
- **Stony (S-type)**, which are made up of silicates and nickel-iron.

Asteroid Bennu, image taken by OSIRIS-REx

Space agencies have sent spacecraft to collect samples from a few asteroids and return them to Earth. These samples can help us better understand the formation of our early Solar System. NASA's OSIRIS-REx spacecraft touched down on an asteroid called Bennu in October 2020 and left in May 2021 to deliver samples of Bennu to Earth. The samples arrived in September 2023, and soon afterward the spacecraft began a new mission as OSIRIS-APophis EXplorer (OSIRIS-APEX) and headed toward an asteroid called Apophis.

Like comets, asteroids can sometimes be knocked out of their orbits. In the asteroid belt, their orbit may be influenced by Jupiter's massive gravity. In the Kuiper Belt, they may be disturbed by interacting with other asteroids.

Asteroids known as centaurs travel in between the orbits of Jupiter and Neptune. Some of them may eventually be ejected from the Solar System, or travel past the inner planets (Mercury, Venus, Mars and Earth), or even burn up if they get too close to the Sun.

Asteroids that cross Earth's orbit are known as **potentially hazardous asteroids (PHAs)** or **near-Earth objects (NEOs)**. Asteroid impacts are a serious business. Sixty-six million years ago, an asteroid roughly 10 kilometers (6 miles) wide slammed into the area that is now the Yucatan Peninsula of Mexico. This event, sometimes referred to as the Cretaceous-Paleogene extinction, caused 75 percent of all animals on Earth to die out, including the dinosaurs.

While our planet does still exist in a shooting gallery (there are billions of objects in the Solar System, ranging from dust grains to kilometer-sized rocks), there are many organizations — including NASA and the European Space Agency — that are constantly searching for any objects that could cross Earth's orbit. To date, NASA says that it has identified nearly 90 percent of any objects 1 kilometer (0.6 miles) wide or larger. And the good news is that we currently know of no PHAs that are a threat to life on Earth up until the year 2200.

But that doesn't mean scientists aren't trying to prepare for the unexpected.

In the early morning hours of February 15, 2013, the sky over Chelyabinsk, Russia, became awash in light. A previously undetected object roughly 20 meters (66 feet) wide entered Earth's atmosphere and exploded in the sky, producing an airburst that shattered windows and injured more than 1,000 people.

Scientists are looking for ways to better detect these somewhat smaller objects, which may not be planet-destroying but could cause harm to populated areas. And they're even looking for ways to change their orbits.

In 2021, NASA conducted the first experiment trying to do just that. It launched its Double Asteroid Redirection Test, or DART. The small spacecraft traveled to a **binary asteroid** — a small asteroid called Dimorphos orbiting a larger asteroid called Didymos. The object of the test was to smash a spacecraft into Dimorphos to see if the impact could change its orbit. If humans can use a spacecraft to alter a PHA's orbit, it will no longer pose a threat to Earth.

The mission was a success. Before the impact, it took Dimorphos 11 hours and 55 minutes to orbit Didymos; after the impact, it took 11 hours and 23 minutes. Photographs by Hubble even showed that Dimorphos formed twin tails for some time after the impact (see the photo on the opposite page).

A **meteoroid** (from the Greek *meteoros*, meaning "high in the air") is a small piece of debris that has broken off from a larger object, usually an asteroid or a comet. Meteoroids are usually the size of a small pebble, but they can be smaller or a little larger. A **meteor** is the light, heat and (occasionally) sound phenomena produced when a meteoroid, collides with molecules in Earth's upper atmosphere. We also call meteors **shooting stars**.

When a meteoroid enters Earth's atmosphere, the surface of the object is heated. Then, at a height typically between 119.8 and 79.9 kilometers (74.5 miles and 49.7 miles), the meteoroid begins to **ablate**, or lose mass.

The vapor trail left after an object entered Earth's atmosphere over Chelyabinsk on February 15, 2013

Meteoroid ablation usually happens through vaporization, although some melting and breaking apart can also occur. If a meteoroid is large enough that it doesn't burn up entirely and reaches the ground, it is called a **meteorite**.

Meteoroids can be divided into two groups: **stream** and **sporadic meteoroids**. Stream meteoroids have orbits around the Sun that often can be linked to a parent object — in most cases a comet, but it can sometimes be an asteroid. When Earth travels within the orbit of a stream, we get a **meteor shower**. The area from which the meteors appear to be originating is called the **radiant**. The constellation containing the radiant (or in some cases a nearby star) is what gives a meteor its name, such as the Perseids or Geminids (two of the most active meteor showers).

Sporadic meteors occur in random parts of the sky, with no radiant.

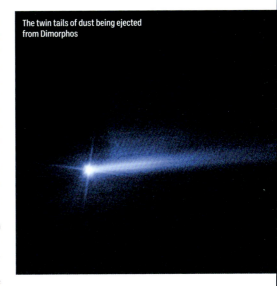

The twin tails of dust being ejected from Dimorphos

A meteor flashes across the sky during the Perseids meteor shower

Meteor Showers in 2026

There is nothing more wonderful than stepping out and looking up at the night sky, only to see a brief streak of light flash against the stars.

Almost monthly, we get major meteor showers. Some showers are best seen from the Northern Hemisphere, while some are better visible in the south, depending on the radiant.

When astronomers talk about the "peak" of a meteor shower, they're referring to the **Zenithal Hourly Rate**, or the ZHR. This is the rate of meteors a shower would produce per hour under clear, dark skies and with the radiant at the zenith. In practice, a single observer will likely see considerably fewer meteors than the ZHR suggests, owing to the presence of moonlight, light pollution, the location of the radiant, poor night vision and inattentiveness.

Here is a list of major meteor showers that you can enjoy simply by staying warm and looking up. No special equipment is needed beyond your eyes; just make sure to give yourself about 20 minutes to adjust to the darkness before searching for meteors.

Quadrantids: December 26, 2025–January 16, 2026

The Quadrantids might be one of the best meteor showers of the year, with a ZHR of 120 for a brief interval of time. The only thing holding it back from earning the title is that January tends to be cloudy over North America, and the shower's peak has a brief window of six hours, making the average hourly rate closer to 25. But there's good news: the meteors often produce bright **fireballs**, or bright meteors with a magnitude brighter than –4.0. This shower's radiant lies between Boötes and Draco. Unfortunately, this year the full Moon will interfere with all but the brightest meteors.
Parent Object: Asteroid 2003 EH1
2026 Peak Night: January 3–4

Lyrids: April 16–30, 2026

After a dearth of meteor showers in the months of February and March, April brings us the Lyrids. This shower produces a ZHR of 18, so it's not a particularly strong shower, but the meteors do tend to produce fireballs. For the peak night, the Moon will be at first quarter.
Parent Object: Comet C/1861 G1 (Thatcher)
2026 Peak Night: April 22–23

Eta Aquariids: April 19–May 28, 2026

The Eta Aquariids aren't a stellar show for the Northern Hemisphere, since the radiant rises in early dawn, but they can still produce a ZHR of 20 meteors, if you care to get up early enough. Rather than fireballs, this shower tends to produce long trains through the sky. The Moon will be in its waning gibbous phase, so it will only interfere with pre-dawn meteor viewing.
Parent Object: Comet 1P/Halley
2026 Peak Night: May 5–6

Perseids: July 17–August 26, 2026

For those in the Northern Hemisphere, the Perseids are considered the best show of the year. The weather is warm and there are fewer clouds at this time of year. The shower's ZHR is 100, though rates of 50 to 75 are more commonly seen on the peak night. Fortunately, it's good news for skywatchers: it will be a new Moon, meaning you can enjoy what is likely to be a wonderful show.
Parent Object: Comet 109P/Swift-Tuttle
2026 Peak Night: August 12–13

Orionids: October 2–November 7, 2026

This shower is considered medium strength, though it can sometimes surprise us with more activity. On average, the Orionids produce a ZHR of 10 to 20 meteors, though from 2006 to 2009, the shower produced roughly 50 to 75 an hour. The meteors that enter our atmosphere are fast

— roughly 66 kilometers (41 miles) per second — and, as a result, produce long, glowing trains. Fireballs are also possible. The Moon will be in its waxing gibbous phase, meaning that only the brightest meteors will be visible.
Parent Object: Comet 1P/Halley
2026 Peak Night: October 21-22

Southern Taurids:
September 10-November 20, 2026
The Southern Taurids last for two months and have several minor peaks. Though this shower produces a ZHR of only 5, it can produce some fireballs. The shower is stronger in the Southern Hemisphere, though we can also catch a few meteors in the north. The Moon will be a waning crescent during the peak night, meaning it won't interfere with meteor watching.
Parent Object: Comet 2P/Encke
2026 Peak Night: November 5-6

Northern Taurids:
October 20-December 10, 2026
Like the Southern Taurids, the Northern Taurids last roughly two months and have a ZHR of 5. There are often reports of an increase of fireballs during the period when the two showers overlap. The waxing crescent Moon will set in the early evening, which is great news for skywatchers.
Parent Object: Comet 2P/Encke
2026 Peak Night: November 12-13

Leonids: November 6-30, 2026
While the Leonids don't produce a high number of meteors an hour (they have a ZHR of 15), they can produce **outbursts**, or meteor storms; the most recent such outbursts occurred in 1999 and 2001. Unfortunately, the next outburst isn't expected until 2099. Still, the Leonids do put on a show, with bright meteors and long-lasting trains. The Moon will be in first quarter, which could interfere with some of the fainter meteors.
Parent Object: Comet 55P/Tempel-Tuttle
2026 Peak Night: November 17-18

Geminids: December 4-17, 2026
Due to the weather — with increasing cloudiness and chilly temperatures — December isn't an ideal time to catch a meteor shower. But December is the month with the most active meteor shower of the year, called the Geminids. This shower has a ZHR of 150, and the meteors it produces are usually bright and colorful. The Moon will be a thin waxing crescent that sets early, so it will not interfere with the show.
Parent Object: Asteroid 3200 Phaethon
2026 Peak Night: December 13-14

Ursids: December 17-26, 2026
Right on the heels of the Geminids is the often-forgotten Ursids meteor shower. The ZHR for this shower is 10, but sometimes you might catch an outburst that could produce a ZHR of 25. The Moon will be in its waxing gibbous phase, meaning that it will interfere with all but the brightest meteors.
Parent Object: Comet 8P/Tuttle
2026 Peak Night: December 22-23

The Moon

The Moon is most often the first astronomical object to capture anyone's attention in the night sky. It's also likely the first object most people see up close, whether it be through binoculars or a telescope.

Our nearest celestial neighbor has been an object of fascination since humans first looked up to the sky. Early scientists in ancient observatories carefully tracked the cycles of the Moon. Some 23 centuries ago, Aristarchus of Samos carefully observed a total lunar eclipse and derived impressively accurate measurements of the Moon's diameter. He estimated its diameter was one-third that of Earth. He was close: the actual percentage is 27.2, and its diameter is roughly 3,540 kilometers (2,200 miles).

It takes about 29.5 days for the Moon to go through its cycle of **phases**. When you can't see the Moon at all, it's called a **new Moon**.

When the Moon is fully illuminated, it is a **full Moon**. The list of phases is new Moon, waxing crescent Moon (when the Moon is just a thin crescent), first quarter Moon (the Moon's eastern half), waxing gibbous Moon (when the Moon is between half-full and full), and then full Moon. The process runs in reverse from full to new: waning gibbous Moon, third or last quarter Moon (the Moon's western half), waning crescent Moon, and new Moon.

The Moon is tidally locked with Earth, meaning that we only ever get to see one face of it. The Sun does shine on the other side of the Moon for some time during every month, so don't call the far side of the Moon the dark side.

While it may seem that we always see the same Moon features month after month, that's not entirely true. The Moon oscillates from our perspective, a process called **lunar libration**. As a result, at certain times we see small zones along the Moon's edge that are normally hidden.

Observing the Moon

You don't need a telescope to enjoy the Moon. If observed even briefly on a regular basis, Earth's only natural satellite has much to offer the naked-eye observer. Keep an eye on it, and you will notice things like its wandering path through the constellations, the changing phases, frequent **conjunctions** (where two objects appear close together in the sky) with planets or bright stars, occasional eclipses, lunar libration, earthshine (illumination on the Moon reflected from our planet) and other wonderful atmospheric effects.

If you happen to take a look through a telescope, binoculars or even a camera, our closest astronomical target offers an amazing amount of interesting detail. There are countless features on the lunar nearside, and more than 1,000 have been formally named by the IAU.

Many of the greatest names in astronomy, exploration and discovery are commemorated through these names. For extra dramatic effect, try looking at a lunar feature when the **terminator** — the line between darkness and sunlight — runs nearby the feature. The extra shadow could make it easier to see details in the object you're interested in looking at.

An estimated three to four billion years ago, our Solar System was bombarded by debris for hundreds of millions of years. This period is called the Late Heavy Bombardment, and you can see many of the scars from that time left over on the Moon.

Because the Moon has almost no atmosphere and no tectonic activity, those scars have remained over billions of years. Some of that activity created deep **impact basins** that filled with lava flows and hardened into basaltic rock. These largely circular regions were termed **maria**, or seas, by early astronomers because they resembled Earth's oceans.

The brighter white areas, which are highlands that surround the maria and dominate the southern portion of the Earth-facing hemisphere, feature many ancient **impact craters**.

After the bombardment slowed to a trickle, the Moon remained relatively free from intense impact activity — meaning the Moon has changed very little since billions of years ago. But, just like on Earth, meteorites still periodically slam into the Moon today.

The Moon map on the next two pages shows a few interesting features for you to target as you observe the Moon.

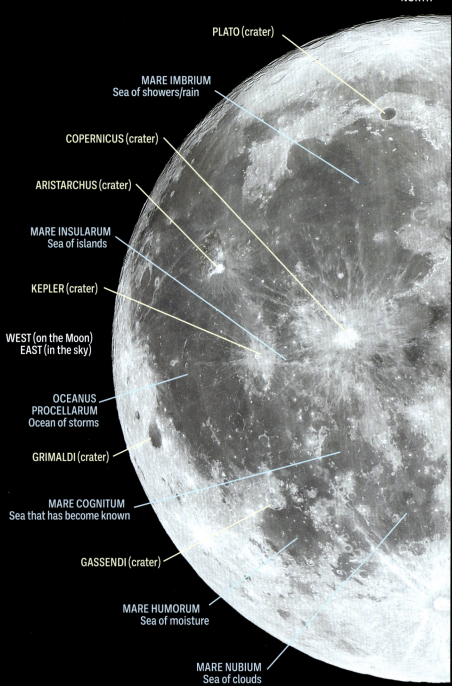

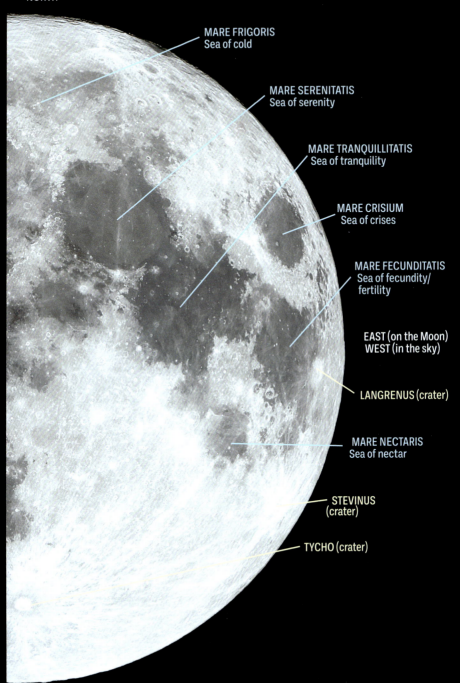

The Sun

Our Sun highly influences Earth. It is connected with our seasons, weather, ocean currents and climate, and its power makes life itself possible.

Unlike Earth, the Sun isn't a solid body, and due to that, different parts of it rotate at different rates. At the equator, for example, it spins roughly once every 25 days, while the polar regions take about 30 days to complete a rotation.

In the core of the Sun, where hydrogen atoms fuse to make helium, temperatures are a searing 15 million degrees Celsius (27 million degrees Fahrenheit). The tremendous amount of energy generated at the core — **thermonuclear fusion** — is carried outward by radiation,

Take Care of Your Eyesight

Do not observe the Sun without special protective equipment, such as a certified solar filter that covers your eyes or telescope aperture entirely. Unprotected observations even for a few seconds could damage your eyesight permanently.

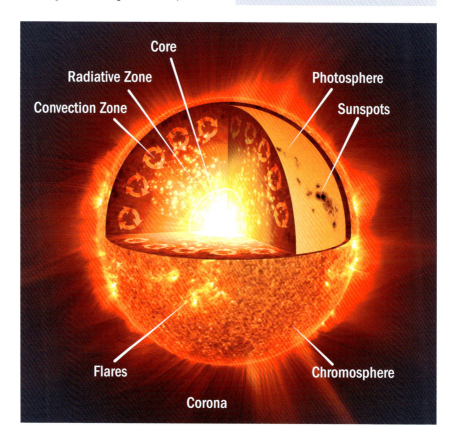

taking roughly 170,000 years to get from the core to the top of the convective/convection zone. As hot as the core is, at the surface, it's a much different story — the temperature is a much "cooler" 5,500 degrees Celsius (10,000 degrees Fahrenheit).

The Sun also has a roughly 11-year cycle, that has a **maximum** (a period of increased solar activity) and a **minimum** (a period of low solar activity). During a maximum, the number of **sunspots**, which are cooler regions on the surface of the Sun, increases. Sometimes, the magnetic field lines of these sunspots can become entangled, finally snapping and releasing a tremendous amount of radiation into space. This process is called a **solar flare**. And often a **coronal mass ejection (CME)** can follow a flare, with charged particles of the Sun speeding outward into space. If these particles reach Earth, they can disrupt radio transmissions, damage satellites and, more positively, interact with our magnetic field. When the particles do interact with the field, they can produce an aurora.

Though beautiful, it must be said that these outbursts from the Sun can also cause power outages, as was experienced in Quebec in 1989. As such, astronomers are keen to better understand our nearest star using many spacecraft — such as the Parker Solar Probe, which was launched in 2018. Keep watching the eight-year adventure of the Parker Solar Probe as it studies the Sun's activity.

A New Solar Cycle

At the end of 2020, we started Solar Cycle 25. After a productive solar maximum in 2024, we can now expect to see solar activity slowly taper off.

The Northern Lights

Observing the Sun

The Sun, our closest star, is a wonderful object to observe with any type of telescope, if used safely. The only safe filter is the kind that covers the full aperture of the telescope, at the front, allowing only 1/100,000th of the sunlight through the telescope. Without this filter, you risk permanently damaging your telescope or, far worse, your eyes. Ensure that the filter material is specifically certified as suitable for solar observing, and *take great caution every time you observe the Sun.*

The preferred solar filters are made of Thousand Oaks glass or Baader film. Thousand Oaks glass gives the Sun a golden-orange color, while Baader film gives the Sun a more natural white look, which can provide good contrast if there are any bright spots, or **plages**, on the Sun. Other solar-filter options include lightweight Mylar film (which gives the Sun a blue-white tint), and eyepiece projection. Projecting the image from the eyepiece onto white cardstock paper, or a wall, produces an image that can be safely shared with others.

While it might seem like the Sun is just this boring yellow ball in the sky, there are many things to see on its surface, including **prominences**, **filaments**, flares and sunspots. These phenomena result from the strong magnetism within the Sun, which erupts to the surface.

Prominences are solar plasma ejections, some of which fall back to the Sun in the form of teardrops or loops; against the Sun's disk, prominences viewed from above appear as dark filaments. Sunspots are cooler regions of the Sun that appear dark on the face of the surrounding surface. If the highly magnetic lines of sunspots intertwine, they can snap and cause a solar flare — a bright, sudden eruption of energy that can last from a few minutes to several hours. It should be noted that to see prominences and some other solar features, you need very specialized filters.

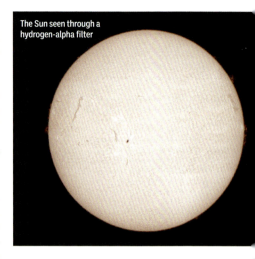

The Sun seen through a hydrogen-alpha filter

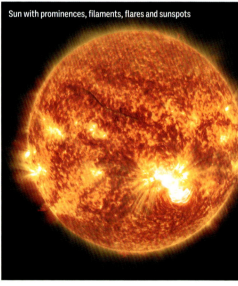

Sun with prominences, filaments, flares and sunspots

If you do not have the proper equipment to observe the Sun, you can always visit the Solar and Heliospheric Observatory (SOHO) and the Solar Dynamics Observatory (SDO) websites (see the list of resources on page 128) to see daily images of the Sun.

Eclipses

Eclipses — whether they're solar or lunar — are seemingly magical occurrences. In fact, that's exactly what many ancient civilizations believed. Several legends suggested a celestial being devoured the Sun during a total solar eclipse. For the Chinese, that being was either a dog or a dragon. The Vikings believed it was two wolves called Hati and Skoll. For the Vietnamese, it was a giant frog. Today, we understand that eclipses are awe-inspiring celestial events resulting from the movement and positions of the Earth, Moon and Sun. There can be as many as seven combined eclipses (i.e., solar and lunar) in a year, but no fewer than four.

A solar eclipse is truly a chance occurrence. The Sun's diameter is roughly 400 times that of the Moon, but the Sun is also about 400 times farther away from Earth than the Moon is from Earth. This means the Moon and the Sun appear roughly the same size in the sky, with only about 1/10th of a degree across of difference. In a **total solar eclipse**, the new Moon covers the entire face of the Sun, revealing the Sun's stunning corona and prominences. As totality begins and ends, sunlight peeking around the mountainous limb of the Moon creates fleeting effects such as the Diamond Ring and Baily's Beads. If the Moon were farther away or smaller, we wouldn't get this marvelous sight. In fact, the elliptical nature of both the Earth's orbit around the Sun and the Moon's orbit around the Earth means that the apparent diameters of both objects

A total solar eclipse

vary significantly. Consequently, it is common for eclipse durations to vary by several minutes.

It should be noted that the totality only occurs in a very narrow path. Outside of this path, observers will only see a partial eclipse. That's why solar eclipse enthusiasts will travel great distances to take in the captivating sight of a totality. Remember, do not look at a solar eclipse unless you use special equipment recommended by The Royal Astronomical Society of Canada or the American Astronomical Society. The only time it is safe to look at a solar eclipse with the naked eye is when it is at its total phase, with the Sun completely covered by the Moon.

Because the Moon orbits at a 5-degree inclination to the ecliptic, we don't get a solar or lunar eclipse every month. Every year, there are two 36-day eclipse seasons during which either solar or lunar eclipses can occur, and the eclipse seasons move backward in time every year by 19 days. Also, because the Moon's orbit is elliptical, the angular size of the Moon varies along its path, and we sometimes have an **annular solar eclipse** (or "ring of fire"), when the Moon's disk does not completely cover the Sun's disk. There may be anywhere between zero to two total or annular solar eclipses in a given year. There may also be **partial solar eclipses**, when the Moon's central shadow entirely misses the Earth.

As for lunar eclipses, these occur when Earth is situated directly between the Sun and the full Moon. The Moon drifts through Earth's two shadows: the **penumbra**, which is the fainter, outer shadow, and the deeper shadow that is called the **umbra**. A **penumbral eclipse** is difficult to see with the naked eye as the brightness of the Moon doesn't appear to dim much. But a **partial** or **total lunar eclipse**, when the Moon passes through the umbra, is much more dramatic. During a total lunar eclipse, the Moon can turn an orange-red color, depending on Earth's atmosphere. There may be anywhere between zero to three total lunar eclipses in a given year. Typically, they occur about two weeks before or after a solar eclipse.

While not all eclipses will be visible from North America, you can watch online on sites like SLOOH or The Virtual Telescope Project. You can also see astronomer Fred Espenak's webpage eclipsewise.com for a comprehensive treatment of solar and lunar eclipses.

Eclipses in 2026

This year there are two solar eclipses and two lunar eclipses, as is most common (but not next year!). This year is a mixed bag, with an annular solar eclipse, a total solar eclipse, a total lunar eclipse and a partial lunar eclipse.

All the times listed are in UTC, or Coordinated Universal Time. See page 46 to learn how to calculate your local time from the time shown in UTC. Be sure to use a proper solar filter when you are viewing a solar eclipse in person. And, if you are clouded out, fortunately, many of these events will also be streamed live online.

A total lunar eclipse

Lunar Eclipses
March 3, 2026: Total Lunar Eclipse

This lunar eclipse will be seen across Canada and the U.S. Those in Alaska, and the Canadian territories of Yukon and parts of Nunavut will be able to experience totality from beginning to end. The rest of Canada and the U.S. will experience the beginning of the partial eclipse and part of totality before the Moon sets.

Eclipse times (UTC)	
Partial eclipse begins (U1)	09:49
Totality begins (U2)	11:03
Greatest eclipse	11:33
Totality ends (U3)	12:02
Partial eclipse ends (U4)	13:17

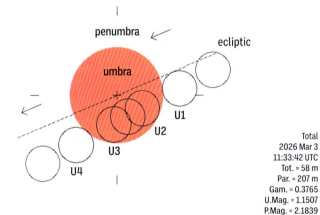

Total
2026 Mar 3
11:33:42 UTC
Tot. = 58 m
Par. = 207 m
Gam. = 0.3765
U.Mag. = 1.1507
P.Mag. = 2.1839

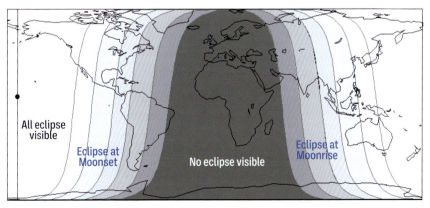

Thousand Year Canon of Lunar Eclipses © 2014 by Fred Espenak

Eclipses 33

August 28, 2026: Partial Lunar Eclipse

The final eclipse of the year is a partial lunar eclipse that will be visible across Canada and the U.S. It will be visible from coast to coast to coast, which means even the northernmost part of Canada will be treated to the show. Though it is a partial eclipse, most of the Moon will be in Earth's shadow, or umbra, with only a sliver in Earth's outer shadow, the penumbra.

Eclipse times (UTC)	
Partial eclipse begins (U1)	02:33
Greatest eclipse	04:12
Partial eclipse ends (U4)	05:52

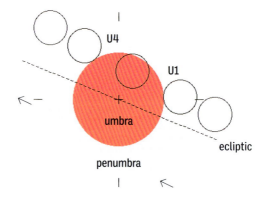

Total
2026 Aug 28
04:12:55 UTC
Par. = 198 m
Gam. = 0.4964
U.Mag. = 0.9299
P.Mag. = 1.9645

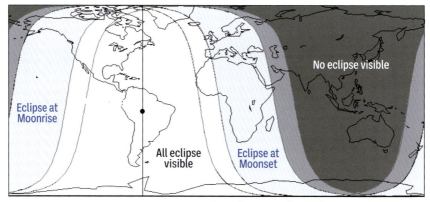

Thousand Year Canon of Lunar Eclipses © 2014 by Fred Espenak

Solar Eclipses

February 17, 2026: Annular Solar Eclipse
The first eclipse of the year is an annular solar eclipse. This type of eclipse occurs when the Moon doesn't cover the disk of the Sun entirely. This is often called the "ring of fire," as the Sun appears as a halo around the Moon. Unfortunately, this eclipse will only be visible in parts of Antarctica. Parts of southern Argentina, Chile and southern Africa will see it as a partial eclipse.

point of greatest eclipse

Thousand Year Canon of Solar Eclipses
© 2014 by Fred Espenak

Total Solar Eclipse: August 12, 2026
Last year, we had only two partial solar eclipses, but this year, we get a total solar eclipse — the first one since the Great North American Eclipse in April 2024. However, this one will only be visible from the Arctic, Greenland, Iceland and Spain. However, northern Canada and the U.S. will see it as a partial solar eclipse.

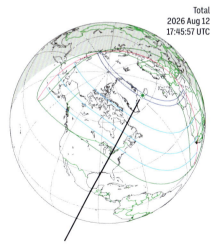

point of greatest eclipse

Thousand Year Canon of Solar Eclipses
© 2014 by Fred Espenak

The Northern Lights

The Sun appears to be a bright, unchanging orb in the sky. However, the Sun is anything but unchanging: As we know, it is a ball of constant activity, and that activity has a large influence here on Earth.

One such activity is a solar flare, which is often followed by a coronal mass ejection (CME) that sends particles speeding along the solar wind. Solar flares can reach Earth after a day or two, and they may disrupt radio transmissions. If Earth is in the path of a CME, the particles can travel down our magnetic field lines toward the poles; this creates the beautiful Northern and Southern Lights, or **aurora borealis** and **aurora australis**, respectively.

As mentioned earlier, the Sun goes through an average 11-year cycle of activity during which it experiences a solar minimum and a solar maximum. In the latter part of 2020, we began a new Solar Cycle, and as the cycle continues, we can expect to see more of the Northern Lights. It's worth noting that over the past few cycles the Sun has been less active than in the past.

Aurorae come in different shapes and sizes. They can be steady, moving or rapidly pulsating. They also come in an array of colors, depending on how the particles interact with molecules at different altitudes. They are also hard to predict and can be difficult to see. Catching them is a special treat, even for experienced astronomers. Be aware that to the unaided eye the auroral colors are usually muted, as the light is not strong enough to stimulate our color vision. The bright colors seen in photographic reproductions of aurorae show up when long exposures are used.

Green is the most common color and occurs when particles interact with oxygen molecules at an altitude of roughly 100 to 300 kilometers (62 to 186 miles). Between 300 to 400 kilometers (186 to 248 miles) the oxygen molecules produce a red color, and below 100 kilometers (62 miles) the particles interact with nitrogen — producing a pink color.

While the interaction of these solar particles can produce a beautiful light show, they can also be destructive.

One of the most powerful events from a CME was called the Carrington Event. In 1859, English astronomers Richard Carrington and Richard Hodgson were the first to ever witness a solar flare. But when the particles reached Earth, they disrupted telegraph systems in North America and Europe, in some reports even setting equipment on fire. The Northern Lights were visible as far south as Honolulu and the Southern Lights as far north as Santiago, Chile.

A massive CME, like the one that caused the Carrington Event, has a real chance of disrupting GPS and satellite communications and destroying electrical grids. As a result, space agencies have been launching more satellites and probes to monitor space weather, and power companies have been developing contingency plans to ensure the next powerful solar eruption doesn't cause such damage.

The Northern Lights over Saskatchewan

The Planets

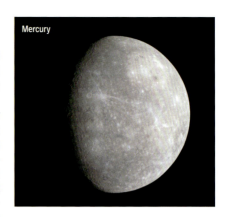
Mercury

As the planets (including Earth) revolve around the Sun, there are a number of notable events that take place, each of which has a specific name and meaning.

The inner planets Mercury and Venus never stray far from the Sun. When either planet is farthest from the Sun in the western evening sky, it is said to be at **greatest eastern elongation**; when either is farthest from the Sun in the eastern morning sky, it is said to be at **greatest western elongation**. These are the best opportunities to view the inner planets, though Venus can be seen at other times.

The term **conjunction** has a specific technical meaning — that is, when two objects have the same right ascension — but amateur astronomers also use the term casually when there is a close approach of two or more celestial objects.

Then there are **oppositions**: when the outer planets (Mars, Jupiter, Saturn, Uranus, Neptune) are opposite the Sun in the sky in right ascension. Around opposition dates, the outer planets appear closer, brighter and larger in diameter — great for telescopic viewing. The Sky Month-by-Month section (pages 46–119) lists eastern and western elongations, conjunctions and oppositions throughout 2026.

Mercury is a place of extremes. As the smallest of our Solar System's eight planets at one-third the size of Earth, Mercury is the closest planet to the Sun, sitting an average distance of 58 million kilometers (36 million miles) away. The planet, however, is only the second-hottest one in our Solar System. Mercury also orbits the Sun faster than any other planet and has the longest solar day, lasting 176 Earth days. Mercury has no moon.

The planet's surface warms to 427 degrees Celsius (801 degrees Fahrenheit) at the peak of its "noonday" heat; on the far side, its temperature drops down to a chilly −163 degrees Celsius (−261 degrees Fahrenheit). Mercury's coldest locations are the bottoms of the perpetually shadowed craters near its poles, where the Mercury Surface, Space Environment, Geochemistry and Ranging (MESSENGER) spacecraft mission to the small planet discovered ice. Temperatures in crater shadows are −183 degrees Celsius (−297 degrees Fahrenheit).

Due to the planet's proximity to the Sun, Mercury is a challenging target to observe. When it is visible in Earth's sky, it appears in a narrow window of time in the dawn twilight just before sunrise or in the evening twilight just after sunset, depending on whether it is west or east of the Sun in the sky. To observe Mercury, one needs a cloud-free horizon, unobstructed by trees or buildings, as it is typically only 15 to 25 degrees away from the Sun when visible.

Venus, the brightest planet in the sky, is a spectacular sight to see. Known as both the "morning star" and the "evening star," Venus is similar in size to Earth. But it's definitely not a place you'd like to visit. The planet is covered in dense clouds and, as a result, suffers from a runaway greenhouse effect. The planet's surface has temperatures of 460 degrees Celsius (860 degrees Fahrenheit) almost constantly.

Venus spins in the opposite direction than the other planets, but very slowly. A day on Venus

A view of the Martian landscape from NASA's Perseverance rover in March 2021

is longer than its year: one day takes 243 Earth days, while one year is 225 Earth days. Its orbit is, on average, 108 million kilometers (67 million miles) from the Sun. Because it is shrouded in white clouds, you won't see surface markings on Venus, but binoculars and small telescopes reveal that Venus passes through phases, similar to the Moon. (Mercury also shows phases, but they are more difficult to see, as the planet's disk appears so small.)

In 2026, Venus will begin to make an appearance low on the western horizon in the middle of February after sunset and will rise higher in the sky until June, when it begins its descent back to the western horizon. It disappears in September and reappears in the eastern sky in November.

Mercury can also be found low on the western horizon in February after sunset. It moves into the morning sky in March, though quite low on the eastern horizon before sunset. It once again returns to the western sky after sunset in June. In August, Mercury can be found low on the eastern horizon. It returns to the western sky in October.

There is no other planet that fascinates humans as much as **Mars**. In 1894, American astronomer Percival Lowell thought he could see deep canals carved through its surface that he believed were evidence of intelligent life. Later observations with spacecraft saw no evidence of these canals.

Mars is the most explored planet in our Solar System, and some even like to joke that it's the only planet with an entirely robotic "civilization" of Earth-made machines.

The orbiters, landers and rovers dispatched to the Red Planet have taught us a lot about its ancient past. Though rocky, dusty and seemingly devoid of any life, some evidence suggests Mars once had an ocean of water covering most of its northern hemisphere. Scientists sometimes cite the existence of water to suggest the planet was once habitable, though we have yet to find proof that life actually ever thrived there.

Mars is about half the size of Earth, with a day very close to Earth's 24-hour day. It lies roughly 228 million kilometers (142 million miles) away from the Sun. Small telescopes (with a 70 mm diameter and below) will show Mars as a featureless red disk; a larger, good-quality telescope at high magnification will show surface features such as ice caps, dark basins and light deserts.

Mars begins to make an appearance in the eastern sky in April just before sunrise. On April 20, four planets meet up in the east before the Sun rises: Mars, Saturn, Mercury and Neptune. Mars remains in the morning sky until December.

The Planets 39

Jupiter is the king of our planetary system, at 11 times wider than Earth. This gas giant boasts the most spectacular storm formation in our Solar System, called the Great Red Spot. Jupiter orbits 772 million kilometers (479 million miles) from the Sun and a single day is about 10 Earth hours long. It orbits the Sun in about 12 Earth years.

Jupiter is the second-brightest planet in our night sky and is also one of the most enjoyable planets to observe. With a modest telescope, you can easily see the cloud bands in its atmosphere. But you can even enjoy the planet through binoculars; if you watch the planet every night, four of its 95 confirmed moons — Io, Europa, Ganymede and Callisto — can be seen changing positions night after night.

The Great Red Spot (GRS) is a massive, swirling egg-shaped storm that has been shrinking for reasons that are unclear to astronomers. In the 1800s, the GRS was estimated at 41,000 kilometers (25,476 miles) along its long axis. NASA's Juno spacecraft measured the GRS at 16,350 kilometers (10,159 miles) in width on April 3, 2017. Several astronomy apps predict the best time on any night for viewing the GRS through a telescope.

Jupiter can be found high in the evening sky in Gemini beginning in January until June, when it dips below the western horizon. It returns to the eastern morning sky in September and remains there until the end of the year.

With its magnificent rings, **Saturn** is often considered the jewel of our Solar System. Through binoculars, Saturn's pancake shape is evident. Look through a telescope, even a modest one, and its fine rings are clearly visible.

Saturn is nine times wider than Earth and 1.4 billion kilometers (886 million miles) away from the Sun. This gaseous planet has the second-shortest day in the Solar System, lasting roughly only 10.7 hours, while its orbit takes 29.4 Earth years. The planet also has more than 274 moons, though its largest, Titan, is the only one easily viewed through a modest telescope.

Saturn is the second-largest planet in the Solar System. It's not the only planet to have rings (Jupiter, Uranus and Neptune all have ring systems), but it has by far the most spectacular ring system we can see. Saturn's rings are made up of billions of pieces of ice that range from tiny dust-sized grains to chunks as big as a house. The ring system is roughly 282,000 kilometers (175,000 miles) across but only about 9 meters (30 feet) tall. There are several rings, mostly close to each other, but the main rings are called A, B, and C. The gap between the outer ring (A) and the first inner ring (B) is called the Cassini Division, following the discovery by Italian astronomer Giovanni Cassini.

Saturn rests along the borders of Pisces and Aquarius on January evenings. It dips to the western horizon in February and returns above the eastern horizon in May. In August the ringed planet will shine in the east in the evening until the end of the year.

Uranus is the faintest of the planets visible to the naked eye (in dark-sky conditions), and

Jupiter

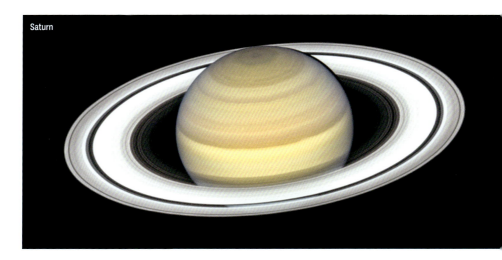

Saturn

the third-largest planet overall. This giant ice planet has 13 faint rings and 28 small moons. But most interestingly, Uranus rotates on its side, at nearly a 90-degree angle.

In 1781, Uranus was the first planet to be discovered with the aid of a telescope by astronomer William Herschel. However, Herschel first believed it was either a star or a comet. It was confirmed as a planet two years later by Johann Elert Bode.

The planet is four times the size of Earth. A day on Uranus takes roughly 17 Earth hours, with one orbit taking 84 Earth years. Uranus orbits on average 2.9 billion kilometers (1.8 billion miles) from the Sun.

Uranus lies in Taurus beginning in January, found high in the east in the evening, until it dips below the western horizon in April. It then swings around to the eastern pre-dawn sky in July. It becomes visible during the evening from September onward.

At an average distance of 4.5 billion kilometers (2.8 billion miles) from the Sun, **Neptune** is the farthest planet in our Solar System, taking about 165 Earth years to complete one orbit. Like Uranus, Neptune is roughly four times larger

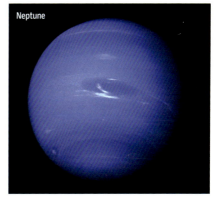

Neptune

than Earth, and it, too, has a faint ring system.

Neptune also holds the distinction as being the windiest planet in our Solar System, with clouds of frozen methane being whipped across the planet at 2,000 kilometers (1,200 miles) per hour. A day on Neptune is about 16 Earth hours long.

Neptune can be found in Pisces in the southwestern sky beginning in January shortly after sunset until March. It makes an appearance in the eastern morning sky in May and then shifts into the evening from August onward.

Deep-Sky Objects

The Universe has many amazing sights to behold, including stunning clusters of stars, nebulae and swirling galaxies.

There are two main types of star clusters: **open** and **globular**. Globular clusters are old star systems at the edge of spiral galaxies that can contain anywhere from thousands to millions of stars, packed in a close, roughly spherical form and held together by gravity.

Two beautiful globular clusters you can see from the Northern Hemisphere include Messier 13, found in the constellation Hercules (a hero from Greek mythology), and Messier 3 in the constellation Canes Venatici (the Hunting Dogs).

Open clusters are found on the **galactic plane**, the plane on which most of a galaxy's mass lies. They contain anything from a dozen to hundreds of stars, but the stars are more spread out than globular clusters. Perhaps the most famous open cluster, the Pleiades (Messier 45) can be spotted with the naked eye and through light pollution.

As well, there are **nebulae** — clouds of dust and gas. These are considered "stellar nurseries," as eventually the gas and dust will coalesce into new stars and potential stellar systems with planets, moons and, possibly, life. One of the most famous and easily visible is the Orion Nebula (Messier 42), found in the winter constellation Orion.

Some nebulae are leftover dust and gas from a **supernova**, an exploding star. There are also **emission nebulae**, which are clouds of interstellar gas excited by nearby stars that emit their own light at optical wavelengths. **Planetary nebulae** are cloudy remnants left over from stars that shed their gas and dust late in their lives. And finally, there are **dark nebulae**, interstellar clouds that are so dense they obscure the light of the objects behind them.

Messier 13, a globular cluster, seen here with Mars (at far left)

Messier Catalogue

The Messier Catalogue is a register of 110 objects in the night sky, including open and globular clusters, nebulae, galaxies and one double star. The catalog was started by Charles Messier in the 18th century. Messier was chiefly interested in finding comets, and the catalog is his list of non-comet objects that he observed. You can find a list of all the objects in the Messier Catalogue on pages 120–125.

Pleiades (Messier 45)

Orion Nebula (Messier 42)

Galaxies

Galaxies come in many different shapes and sizes. There are four main types, however: elliptical, spiral, barred spiral and irregular. The diagram below shows Edwin Hubble's scheme for classifying galaxies, colloquially known as the "tuning fork" because of its shape.

Elliptical galaxies lack structure and look roughly egg-shaped. Shapes range from almost circular (E0) to very elliptical (E7).

Spiral galaxies have both a large central bulge and a thin disk of stars composed of spiral arms. Much like elliptical galaxies, these also range from tight spirals (Sa) to more diffuse (Sd).

Barred spiral galaxies have an elongated central bulge that is surrounded by arms that range from tight (SBa) to loose (SBd).

There are also disk galaxies that don't have spiral arms. These are called **lenticular (lens-shaped) galaxies**. They are classified as S0.

The most common galaxy is the spiral, accounting for more than 75 percent of galaxies in the visible Universe. Our galaxy, the Milky Way, is believed to be a barred spiral. Our closest neighboring spiral galaxy is the Andromeda Galaxy (Messier 31), which is easily visible through binoculars or with the naked eye in dark skies.

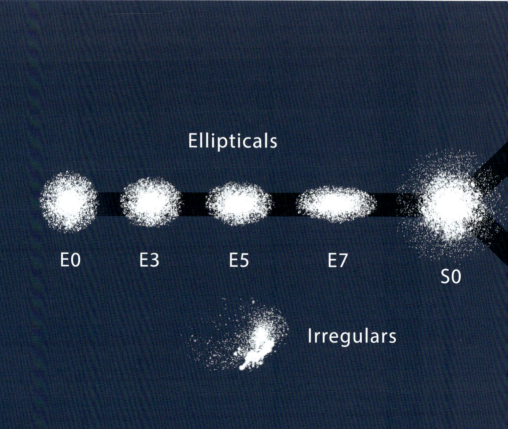

NGC 3949, a galaxy thought to resemble our Milky Way galaxy

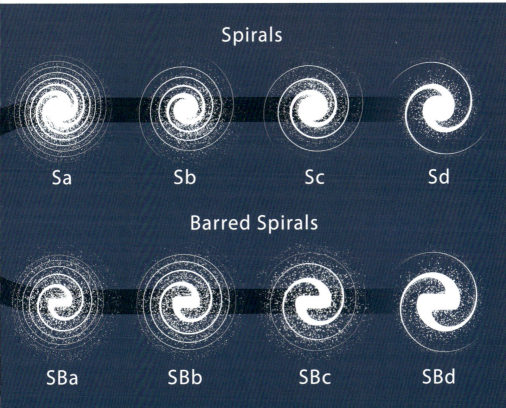

Deep-Sky Objects 45

The Sky Month-by-Month

Introduction

The following pages are your guide to the sky for every month in 2026. Each month features a calendar of events, information about Moon phases, sky charts facing south and north and descriptions of interesting objects to target.

Monthly Events

This section summarizes the events of each month, including Moon phases, conjunctions between the Moon and planets, conjunctions between planets and other planets, eastern and western elongations, meteor shower peaks and much more.

All times are listed in UTC, or Coordinated Universal Time, which is the standard time used by astronomers throughout the year. The map shown below will guide you on how to calculate your local time from the time shown in UTC. It's important to note that most of North America will observe Daylight Saving Time (DST) between March 8, 2026, and November 1, 2026. Between these two dates, you will need to add an additional hour to your local time.

You'll notice some of the times listed fall during daylight, when observation is likely impossible. However, you can use this date

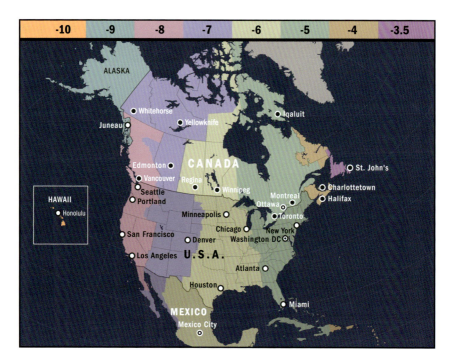

and time as a guide for observing an event on either the preceding night or the following night. Depending on your location, objects might also be below the horizon at the stated time, so again, use this calendar as a guide for which night these events can be observed. Similarly, the coordinates given as the distance between celestial objects may not reflect exactly what you will see in the night sky. This is particularly true for close approaches that take place during the day, local time. For example, the calendar of events might say Mercury is 1.5 degrees south of Jupiter at 21:00 UTC, but by the time Mercury and Jupiter become visible to an observer in Eastern Time living in Toronto, Canada, that separation might have increased to 2.4 degrees.

This section also features a calendar layout with the Moon phases for each day shown, as well as a description of interesting Moon events for the month. You might notice some discrepancies between the coordinates given in these descriptions and those listed in the table of events. Because the Moon moves in right ascension so quickly, conjunctions between the Moon and planets as listed in the table may appear different for North American observers, and the separation between the objects may be more than suggested. The coordinates in the descriptions have been altered to more closely reflect the view from North America, but of course there may be more variation based on when the objects are visible to you in the night sky.

The Moon phases shown in the calendar are based on the day they occur in UTC time. For some observers in North America, that might mean the ideal time to watch, for example, the full Moon rising would be the night before.

Sky Charts and Descriptions

Each month features two sky charts, one facing south and the other facing north. These charts highlight select constellations, stars, clusters, nebulae, planets and galaxies visible in the night sky. While these sky charts are an accurate representation of the sky, it should be noted that factors like light pollution, smoke, cloud cover and so on can affect how much you see, and some of the fainter stars shown on the charts may not be visible.

The charts are drawn at a latitude of 45 degrees north. Observers north or south of this latitude will see slightly more of the northern or southern sky, respectively. The charts show the sky at 10:00 p.m. local standard time on the 15th of each month. You will also have the same view of the sky at 11:00 p.m. local standard time at the beginning of each month and at 9:00 p.m. local standard time at the end of each month. We note these times on the sky charts and also include Daylight Saving Time in parentheses between March and November. The planets shown on the sky charts will move slightly relative to the stars over the course of the month.

The preceding month's charts can be used for sky viewing two hours earlier, and the following month's charts can be used for sky viewing two hours later. So, for example, if you wanted to view the sky at 8:00 p.m. local time in February, you would refer to January's sky chart. If you're referring to a previous or later month's sky chart, note that the positions of the planets will not be correct.

January

January's Events

January always gets off to a start with a bang, thanks to the Quadrantids meteor shower.

This shower's Zenithal Hourly Rate (see p. 22) is over 100 meteors an hour, but only during its unusually short, six-hour window. The shower peaks on the night of January 3 to 4. Unfortunately, this year the Moon will be full, washing out all but the brightest meteors.

Bright Jupiter will be shining in the east after dark, a difficult sight to miss among the stars of Gemini. On January 9, Jupiter will be closest to Earth at roughly 633,000,000 kilometers (393,327,965 miles). Then, on January 10, it will be at opposition, while Earth passes between Jupiter and the Sun.

You can find Saturn low in the southwest after dark near the border between Pisces and Aquarius. Mercury, Mars and Venus are all too close to the Sun to see this month.

As for the two ice giants, Uranus can be found in Taurus, while Neptune can be found in Pisces near Saturn.

Earth will be at perihelion — closest to the Sun in its yearly orbit — on January 3.

Calendar of Events

Day	Time (UTC)	Event
01	21:43	Moon at perigee: 360,300 km (223,880 mi.)
03	10:03	Full Moon
03	16:59	Earth at perihelion: 0.9833 astronomical units (au)
03		Quadrantids meteor shower peak
03	22:01	Jupiter 3.7°S of the Moon
05	01:09	Beehive 1.3°S of the Moon
06	16:20	Regulus 0.5°S of the Moon
10	08:22	Jupiter at opposition
10	15:48	Last quarter of the Moon
10	23:50	Spica 1.8°N of the Moon
13	20:47	Moon at apogee: 405,400 km (251,345 mi.)
14	19:28	Antares 0.6°N of the Moon
18	19:52	New Moon
23	12:31	Saturn 4.5°S of the Moon
26	04:47	First quarter of the Moon
27	21:07	Pleiades 1.1°S of the Moon
29	21:53	Moon at perigee: 365,900 km (227,360 mi.)
31	02:31	Jupiter 3.9°S of the Moon

The Moon This Month

January 2026

SUN	MON	TUES	WED	THURS	FRI	SAT
				1	2	3 Full Moon
4	5	6	7	8	9	10 Last Quarter
11	12	13	14	15	16	17
18 New Moon	19	20	21	22	23	24
25	26 1st Quarter	27	28	29	30	31

On January 1, the Moon will be at perigee, or the closest to Earth in its monthly orbit.

The full Moon occurs just two days later, on January 3, and interestingly it will also be the same day Earth is at perihelion — the closest point to the Sun. There will also be a so-called quasi-conjunction of the Moon and Jupiter on this day. Two objects in the sky appearing close together is called an "appulse" — a true conjunction is when two objects have the same celestial longitude (right ascension).

In the evening on January 4, the still-bright, waning gibbous Moon will shine just above Messier 44, also known as the Beehive Cluster. On the morning of January 14, the pretty crescent Moon will shine near the bright star Antares. On January 22 and 23, the waxing crescent Moon will shine near Neptune and Saturn in the west after sunset.

The last quarter of the Moon happens on January 10, the new Moon on January 18 and the first quarter on January 26.

A stunning view of the waxing gibbous Moon

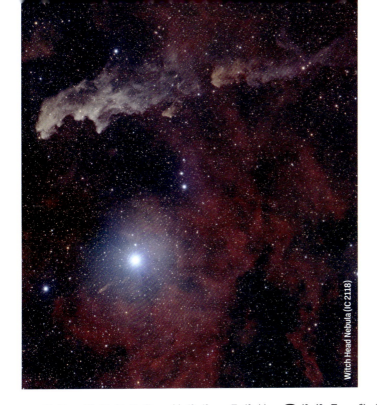

Witch Head Nebula (IC 2118)

January 2026

Highlights in the Southern Sky

We begin in the southeast, where the most prominent constellation in the sky, **Orion (the Hunter)**, provides a multitude of targets for those with telescopes or even just a simple set of binoculars.

His showpiece is the **Orion Nebula (Messier 42)**. This beautiful cloud of gas is visible with the naked eye, but binoculars can help you spot more of the visible gas. Through even a modest telescope, this stellar nursery is bursting with beauty. It has a magnitude of 4 and is one of the brightest nebulae in the night sky. Small telescopes will also reveal the **Trapezium**, a collection of four young, hot stars at the nebula's center.

Barnard's Loop is a large emission nebula and star-forming region that arcs around the stars in **Orion's Belt**. It spans roughly 10 degrees across the sky, covering a fair portion of Orion. Long-exposure photographs reveal it best.

One of the most interesting things about Orion is its stars. You can find **Betelgeuse**, a hot red giant nearing the end of its life at the Hunter's left shoulder. You can compare it to **Rigel**, a bright blue-white supergiant that marks Orion's right foot.

To the right of Rigel, you can find the **Witch Head Nebula (IC 2118)** lying in **Eridanus (the River Eridanus)**. For those with telescopes, this nebula sits 900 light-years from Earth and spans roughly 70 light-years across. Because it is quite a faint nebula, at magnitude 13, it is best seen through large telescopes and makes a stunning photographic target.

To the left of Orion is **Gemini (the Twins)** with its two unmistakable fraternal twin stars, **Castor** and **Pollux**. While Mars visited the constellation in 2025, this year it's the mighty **Jupiter**'s turn. By mid-month, you can find the most massive planet in the Solar System gleaming near **Wasat**, which marks Pollux's waist.

The "star" of the southern sky, you might say, is **Sirius** — the brightest in the night sky. As part of **Canis Major (the Great Dog)**, Sirius is nicknamed the **Dog Star**. It's actually a binary star that lies 8.6 light-years away.

January 51

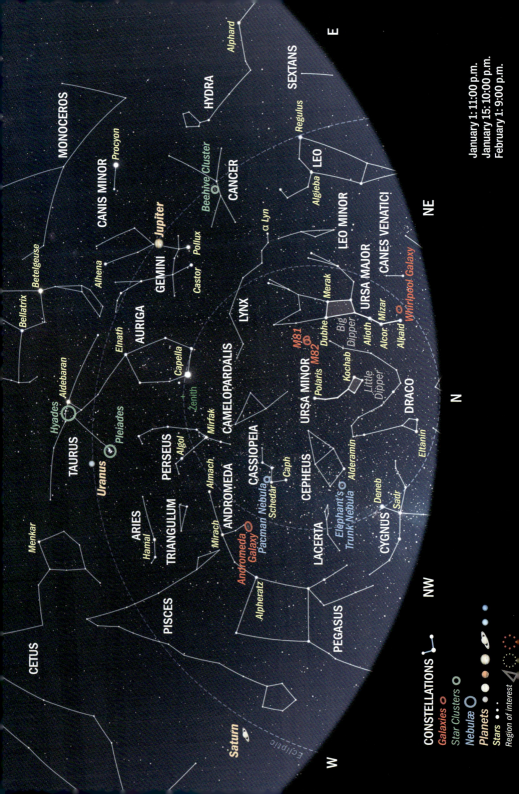

Highlights in the Northern Sky

Since we're looking north, let's start due north with **Ursa Minor (the Little Bear)**, which contains the **Little Dipper** asterism. The most important of stars (especially for those astronomers with equatorial mounts) is **Polaris**, also known as the **North Star**. The lesser-known star **Kochab** is the second brightest in this dim constellation. But, like so many stars we see in the night sky, Kochab is actually a double star with a visual companion star that is much dimmer. Kochab also hosts another interesting object, the exoplanet **Beta Ursae Minoris b**, a gas giant that is roughly 6.1 times more massive than Jupiter.

The most familiar asterism in the night sky for those in the northern hemisphere is the **Big Dipper**, also known as the **Plough**, part of the constellation **Ursa Major (the Great Bear)**. There are a few telescopic treats around this spoon-like asterism, including a few galaxies. Among them are **Bode's Galaxy (Messier 81)** and the **Cigar Galaxy (Messier 82)**. The close-together pair are a favorite target of astrophotographers. They were discovered in 1774 by Johann Elert Bode (which is why they were sometimes referred to as nebulae, as galaxies were yet to be recognized as such). The duo can be seen in medium-sized binoculars in dark skies, mainly as faint fuzzy spots. Telescopes of any size will reveal more detail, particularly for Messier 81.

In the northwest you can also find **Cassiopeia (the Queen)**, a W-shaped formation (or M-shaped, depending on whether it is high or low in the sky). **Schedar** is its brightest star, but the constellation is unmistakable, even in moderately light-polluted skies.

One of Cassiopeia's most interesting nebulae is the **Pacman Nebula (NGC 281)**, named for its resemblance to the main character of the famous video game. The nebula was discovered in 1883 by E.E. Barnard (the namesake of Barnard's Loop found in Orion). It sits approximately 9,500 light-years from Earth and is visible in moderately large telescopes.

Pacman Nebula (NGC 281)

January 2026

February

February's Events

We begin the month with a very special occurrence: a lunar occultation of Regulus.

The occultation falls on the night of February 2, when the Moon will slowly glide in front of Regulus, the brightest star in Leo. About an hour later, Regulus will pop back out from behind the Moon. Though not visible across all of North America, the occultation can be seen from roughly the midwestern U.S. and prairie provinces in Canada across to the eastern seaboard (though it will not be visible in Florida).

Jupiter is still shining brightly in Gemini. Since it's so well placed in the sky, this would be a great time to watch the motion of the giant planet's four largest moons, Io, Europa, Ganymede and Callisto, night after night. Because they orbit Jupiter quite quickly, their positions change every night as seen from Earth.

There's still more to come with the planets. On the nights surrounding February 20, there is a conjunction of Saturn and Neptune low in the southwest. Mercury will be visible low in the western sky after sunset during mid-February, with it being highest in the sky on February 20. Around that date, Venus will begin to appear very low in the west after sunset. Uranus can be found in Taurus.

On February 17, there is an annular solar eclipse that will be visible in Antarctica and parts of southern Africa and southern South America.

Calendar of Events

Day	Time (UTC)	Event
01	11:41	Beehive 1.3°S of the Moon
01	22:09	Full Moon
03	02:48	Regulus 0.4°S of the Moon
07	08:26	Spica 2°N of the Moon
09	12:43	Last quarter of the Moon
10	16:52	Moon at apogee: 404,600 km (251,407 mi.)
11	03:19	Antares 0.8°N of the Moon
17	12:01	New Moon
17	12:13	Annular solar eclipse (not observable from North America)
19	17:59	Mercury 18.1°E of the Sun (greatest elongation east)
19	23:54	Saturn 4.8°S of the Moon
24	02:43	Pleiades 1.2°S of the Moon
24	12:28	First quarter of the Moon
24	23:18	Moon at perigee: 370,100 km (229,969 mi.)
27	06:26	Jupiter 4°S of the Moon
28	20:07	Beehive 1.3°S of the Moon

The Moon This Month

SUN	MON	TUES	WED	THURS	FRI	SAT
1 Full Moon	2	3	4	5	6	7
8	9 Last Quarter	10	11	12	13	14
15	16	17 New Moon	18	19	20	21
22	23	24 1st Quarter	25	26	27	28

As mentioned, a lunar occultation of Regulus occurs on February 2. On February 18, the crescent Moon, which will be only 2.2 percent illuminated, will shine near Mercury above the western horizon after sunset. It could be a nice challenge for those who may want to photograph the conjunction. The Moon will occult Mercury for observers in parts of Mexico, but the event will not be visible from Canada or the U.S.

On February 23, the half-illuminated Moon will brush the Pleiades star cluster. On February 26, there will be a conjunction of the waxing gibbous Moon with Jupiter. On February 28, the bright Moon shines near Messier 44, the Beehive Cluster.

The full Moon falls on February 1, with the new Moon falling on February 17. It will be in its last-quarter phase on February 9 and its first-quarter phase on February 24.

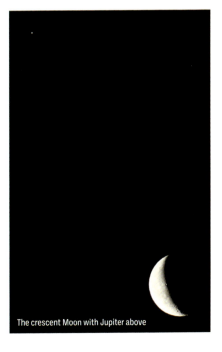

The crescent Moon with Jupiter above

February 55

February 2026

Beehive Cluster (Messier 44)

Highlights in the Southern Sky

Starting in the east, you'll find **Leo (the Lion)**. The constellation is fairly easy to spot. The brightest star in the constellation is **Regulus**, with **Denebola** as the second brightest marking the lion's tail. Owners of backyard telescopes will want to view the **Leo Triplet**, a triangular group of interacting spiral galaxies named **Messier 65**, **Messier 66** and the **Hamburger Galaxy (NGC 3628)**.

You can also find the beautiful **Beehive Cluster (Messier 44)** in Cancer. This is a gorgeous open cluster containing at least 1,000 stars, lying 577 light-years from Earth, making it one of the nearest open clusters. Though visible to the unaided eye under dark skies, binoculars and telescopes bring its stars to life.

Orion (the Hunter) dominates the southwestern sky. Look to Orion's upper right for the **Hyades (Melotte 25)** in **Taurus (the Bull)**. This stunning open cluster is a treat for binoculars, containing hundreds of stars. It's easy to spot as it looks like the letter "V" anchored by the star **Aldebaran**, which isn't part of the cluster but happens to be in the line of sight and closer to Earth than the cluster itself.

Near Hyades is yet another beautiful open cluster, the **Pleiades (Messier 45)**, also known as the **Seven Sisters**, **Subaru** and **Matariki**. This cluster contains hot, blue and extremely luminous B-type stars (see page 15). Like the Beehive, it is one of the nearest clusters to Earth, at just over 444 light-years. The cluster is easy to spot in even moderately dark skies, appearing as a fuzzy patch. Photographs reveal blue haloes, bright starlight reflected off interstellar dust.

February 57

Highlights in the Northern Sky

In the northwest lies **Perseus (the Hero)**, with its brightest star **Mirfak**. The star is an F-class supergiant (see p. 15) that is roughly 65 times larger than our Sun. It is also a variable star in that it changes in brightness over time; however, the change is quite minor compared to other variable stars. Mirfak is surrounded by a gorgeous cluster of stars best seen in binoculars.

Just below **Miram**, the seventh-brightest star in the constellation, you can find the **Double Cluster (NGC 869 and NGC 884)**, which makes another nice binocular target. Those stars lie roughly 7,500 light-years from Earth.

Directly above Perseus is **Auriga (the Charioteer)**, with its extremely bright star **Capella**. In between **Elnath** and **Al Kab**, two of Auriga's stars, is the **Flaming Star Nebula (IC 405)**. You can also find the **Pinwheel Cluster (Messier 36)**, a nice open cluster of stars, above Elnath.

Canes Venatici (the Hunting Dogs), which represents the hunting dogs of **Boötes (the Herdsman)** is climbing the eastern sky. Most would find it difficult to spot Canes Venatici in the night sky because it has only two bright stars, **Cor Caroli** and **Chara**. However, it is a treasure trove of objects, including the well-known **Whirlpool Galaxy (Messier 51)**, found just off the handle of **Ursa Major**'s star **Alkaid**. You can also find the **Sunflower Galaxy (Messier 63)**, the **Cat's Eye Galaxy (Messier 94)** and the **Cocoon Galaxy (NGC 4490)** just above Chara.

Cat's Eye Galaxy (Messier 94)

March

March's Events

March evenings host both winter and spring constellations. If you're an early riser, you can see the summer constellations, including the Summer Triangle asterism. Bright red Antares is now in the south before dawn, signaling the rise of Sagittarius.

Mid-March is the best time for the yearly Messier Marathon, when astronomers try to view all 110 Messier objects in a single night. Many of these objects can be found using modest binoculars, while others require a telescope.

Mercury, Neptune and Saturn have now all sunk below the western horizon by the evening, but Venus will be climbing the western sky after sunset. Uranus can be found in Taurus all evening, just below the Pleiades. Meanwhile, Jupiter is an all-night target in Gemini. It ends its westerly retrograde motion on March 10.

There will be two planetary conjunctions in the western sky this month. On March 6 and 7, Venus, Saturn and Neptune meet up very low in the western sky, near the horizon, but Neptune will be almost impossible to see. On March 7 and 8, Saturn and Venus will share the view in a telescope after sunset.

This month we get a special treat. Before dawn on March 3, there will be a total lunar eclipse visible across Canada and the U.S.

Calendar of Events

Day	Time (UTC)	Event
02	12:00	Regulus 0.4°S of the Moon
03	11:35	Total lunar eclipse
03	11:38	Full Moon
06	17:24	Spica 2°N of the Moon
10	11:32	Antares 0.8°N of the Moon
10	13:43	Moon at apogee: 404,400 km (251,283 mi.)
11	09:39	Last quarter of the Moon
15	19:13	Mars 3.4°N of the Moon
19	01:23	New Moon
20	12:39	Venus 4.7°S of the Moon
20	14:46	Vernal (spring) equinox
22	11:40	Moon at perigee: 366,900 km (227,981 mi.)
23	08:32	Pleiades 1.1°S of the Moon
25	19:18	First quarter of the Moon
26	12:13	Jupiter 4°S of the Moon
28	02:15	Beehive 1.3°S of the Moon
29	19:00	Regulus 0.4°S of the Moon

The Moon This Month

March 2026

SUN	MON	TUES	WED	THURS	FRI	SAT
1	2	3 Full Moon	4	5	6	7
8	9	10	11 Last Quarter	12	13	14
15	16	17	18	19 New Moon	20	21
22	23	24	25 1st Quarter	26	27	28
29	30	31				

The Moon will shine just below Antares at sunrise on March 10. The Moon will shine close to Venus in the western sky on March 19 and 20, but you may be hard-pressed to notice its very thin crescent. Knowing where the Moon is may provide an opportunity to photograph the meeting, but it will be a challenge.

On March 22, the Moon will pass close to the Pleiades (Messier 45) and Uranus. The Moon will be a waxing crescent and 20 percent illuminated.

A few days later, on March 25 and 26, there will be a conjunction of the Moon and Jupiter. By that point, the Moon will be waxing gibbous and roughly 64 percent illuminated.

The full Moon falls on March 3 — when we will experience a total lunar eclipse in the very early morning. The last quarter will occur on March 11, the new Moon is on March 19 (best viewed on the night of March 18 to 19) and the first quarter is on March 25.

Total lunar eclipse

March 61

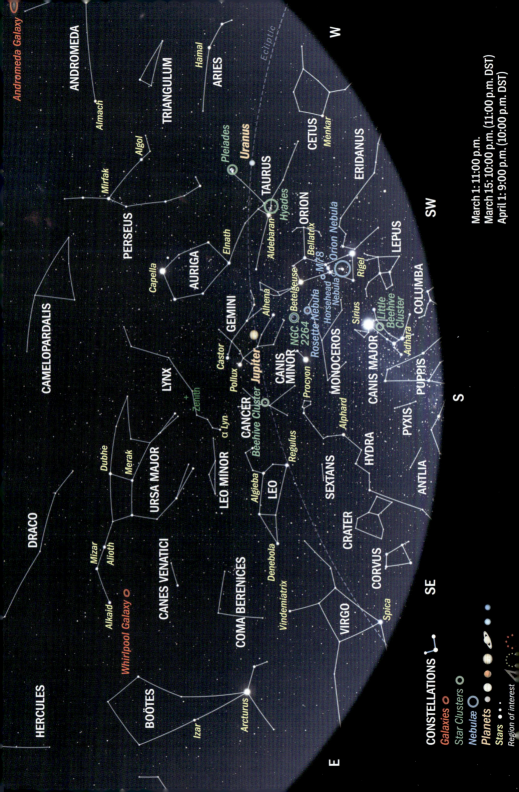

Horsehead Nebula (Barnard 33)

Highlights in the Southern Sky

Sirius is quite prominent in the south as part of **Canis Major (the Great Dog)**. Sirius, the brightest star in the night sky has a magnitude of −1.46. It is so bright because it's 25.4 times more luminous than the Sun and relatively close to us. Like many stars in our night sky, Sirius is part of a binary star system.

Though low on the horizon, you can find a nice cluster in Canis Major below Sirius. The **Little Beehive Cluster (Messier 41)** is a bright open cluster that lies roughly 2,300 light-years from Earth. It can be found in the same binocular field of view as Sirius.

Above Sirius, in **Monoceros (the Unicorn)**, you can find the **Rosette Nebula (NGC 2237)**, a large emission nebula. Its center hosts an open cluster that is believed to have been formed in the last 5 million years. Those stars are causing the gas in the nebula to glow.

Above the Rosette Nebula lies the **Cone Nebula**, a pillar of gas and dust that lies roughly 2,700 light-years from Earth. It is part of a large star-forming region that also hosts the **Christmas Tree Cluster** and the **Fox Fur Nebula**. Together they all share the designation **NGC 2264** in the New General Catalogue.

Orion (the Hunter) is still present, now in the southwestern sky. There are a couple of nebulae near the belt stars of Orion that are stunning to view and photograph. One is **Messier 78**, a reflection nebula that lies about 1,600 light-years from Earth. Though it can be seen through 10 × 50 binoculars at a dark-sky site, it's best seen with a moderately large telescope.

Nearby is the **Horsehead Nebula**, also known as **Barnard 33**, which is located roughly 1,375 light-years from Earth. The horse's head is made up of swirling clouds of dust and gas that are blocking the hydrogen gas behind the nebula. You'll need a large telescope to see it or if you're an astrophotographer, you'll need to take a long-exposure photograph.

Taurus (the Bull) is also in the southern sky with its beautiful open clusters, **Hyades (Melotte 25)** and the **Pleiades (Messier 45)**.

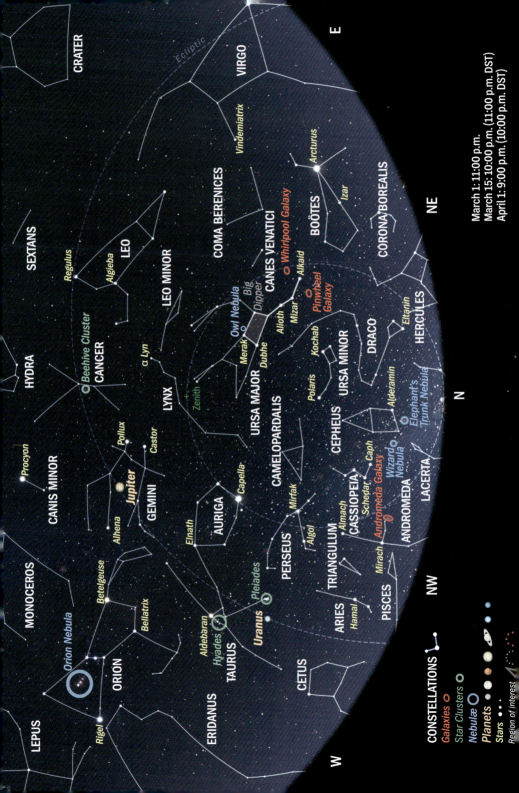

March 2026

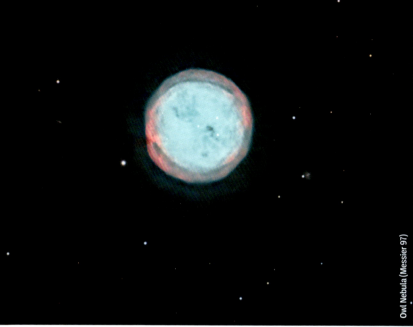

Owl Nebula (Messier 97)

Highlights in the Northern Sky

Boötes (the Herdsman), with its unmistakable star **Arcturus** shining brightly, is beginning to inch its way up above the horizon in the northeast. **Ursa Major (the Great Bear)** is high in the northeastern sky. Find **Alkaid**, the first star in the **Big Dipper**'s handle and that will lead you to the **Pinwheel Galaxy (Messier 101)**. This popular photographic target lies roughly 21 million light-years from Earth. Though the galaxy is quite large, its surface brightness is low, making it a somewhat challenging target visually.

Still sticking with Ursa Major, you can find the round **Owl Nebula (Messier 97)** near **Merak**, the bottommost star in the bowl of the Big Dipper.

Meanwhile, **Draco (the Dragon)** is slithering its way between Ursa Major, **Ursa Minor (the Little Bear)** and Boötes. **Cassiopeia (the Queen)** is sinking lower in the northwestern sky, as is **Perseus (the Hero)** above her.

March 65

April

April's Events

In April Mercury, Mars, Saturn and Neptune are all above the eastern horizon before sunrise, though not yet easy to see. Venus is dominating the western sky after sunset. Jupiter is shining brightly in Gemini high in the west.

On April 20, Mars, Saturn and Mercury form a line just above the eastern horizon before sunrise, but you'll need to use binoculars to catch sight of them. (For eye safety, turn binoculars away from the east before the Sun rises.)

Uranus and Venus shine near the Pleiades low in the west after sunset on April 23 and 24. Binoculars will show them all at once.

The Lyrids meteor shower peaks on the night of April 22 to 23. This shower doesn't often produce long trains, but it can include some fireballs. For the peak night, the Moon will be at first quarter.

Calendar of Events

Day	Time (UTC)	Event
02	02:12	Full Moon
03	01:32	Spica 1.9°N of the Moon
03	22:59	Mercury 27.8°W of the Sun (greatest elongation west)
06	19:21	Antares 0.7°N of the Moon
07	08:32	Moon at apogee: 405,000 km (251,655 mi.)
10	04:52	Last quarter of the Moon
16	00:45	Mars 3.8°S of the Moon
17	11:52	New Moon
19	06:57	Moon at perigee: 361,600 km (224,688 mi.)
19	08:48	Venus 4.7°S of the Moon
19	16:28	Pleiades 1°S of the Moon
19	19:04	Saturn 1.2°N of the Moon
20	09:41	Saturn 0.5°N of Mercury
20	22:15	Mars 1.7°N of Mercury
22		Lyrids meteor shower peak
22	22:06	Jupiter 3.6°S of the Moon
24	02:32	First quarter of the Moon
24	04:15	Pleiades 3.5°S of Venus
24	07:46	Beehive 1.1°S of the Moon
26	00:37	Regulus 0.2°S of the Moon
30	08:17	Spica 2°N of the Moon

The Moon This Month

April 2026

SUN	MON	TUES	WED	THURS	FRI	SAT
			1	2 Full Moon	3	4
5	6	7	8	9	10 Last Quarter	11
12	13	14	15	16	17 New Moon	18
19	20	21	22	23	24 1st Quarter	25
26	27	28	29	30		

Before sunrise on April 15, the pretty, waning crescent Moon will shine above the eastern horizon over Saturn, Mars, Neptune and Mercury. Observers located at southerly latitudes will have the best odds of seeing the event. You'll need an unencumbered view of the eastern horizon and binoculars to catch the show.

On April 18, the waxing crescent Moon will shine near Venus and Uranus in the western sky after sunset. And a few days later, on April 22, the Moon will meet Jupiter in Gemini.

The full Moon falls on April 2, while the last quarter falls on April 10. (These are best viewed the night before, on April 1 and April 9, respectively.) The new Moon occurs on April 17, and the first quarter is on April 24 (but best viewed on April 23).

The waxing crescent Moon with Venus

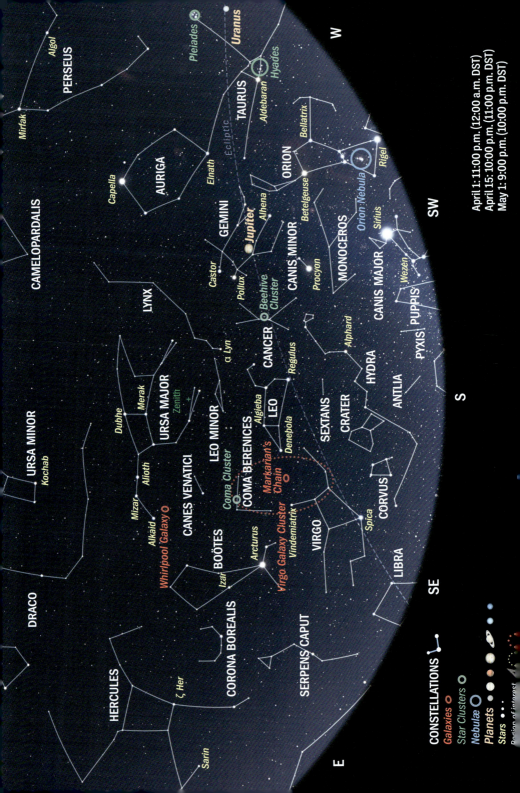

Markarian's Chain

Highlights in the Southern Sky

Virgo (the Maiden) now occupies a large part of the southeastern sky, with its star **Spica** shining brightly. The bluish-white star is the brightest in the constellation and sits roughly 250 light-years from Earth.

One of the most remarkable things about this constellation is the vast collection of galaxies that can be found within it, many of them visible in backyard telescopes. The **Virgo Cluster** is a collection of roughly 2,000 galaxies and includes the **Coma Cluster** of galaxies found in **Coma Berenices (Berenice's Hair)**. This group of galaxies spans roughly 15 million light-years and is part of the larger and more massive **Virgo Supercluster**.

On the border between Virgo and Coma Berenices is **Markarian's Chain**, a collection of a dozen galaxies that can be seen through a moderately large telescope.

Cancer (the Crab) in the southwestern sky hosts the beautiful **Beehive Cluster (Messier 44)**. This swarm of roughly 1,000 stars can be seen with just average binoculars or even your unaided eyes on a very dark night.

Leo (the Lion) now dominates the southern sky with bright **Regulus** taking center stage.

Below Leo and Virgo, you can find **Corvus (the Crow)** along with **Crater (the Cup)**. Lengthy **Hydra (the Water Snake)** winds its way around the pair. **Procyon** is shining brightly in **Canis Minor (the Little Dog)**, while **Castor** and **Pollux**, the brightest stars in **Gemini (the Twins)**, are beginning to descend in the west with Jupiter.

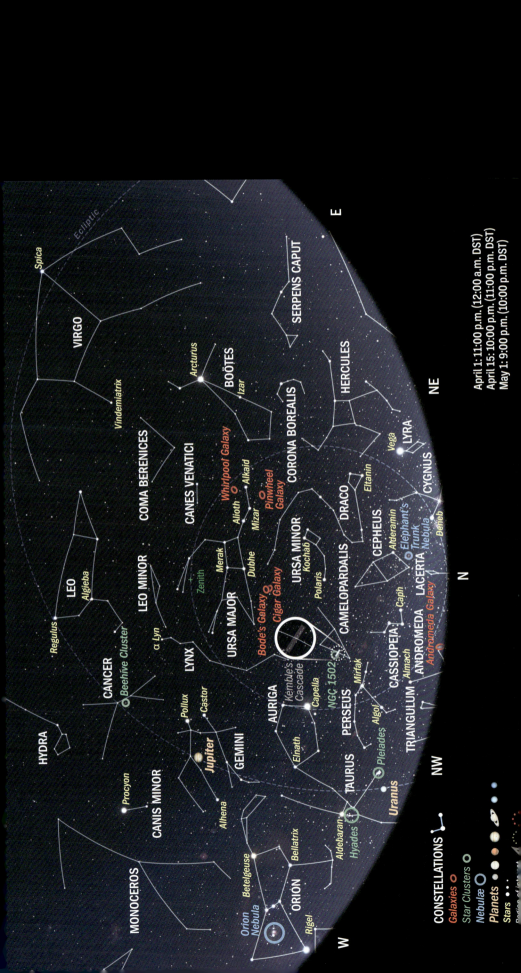

Kemble's Cascade

Highlights in the Northern Sky

Beginning in the northwest, we see **Auriga (the Charioteer)** with its brightest star, **Capella**, shining ever-so brightly as it descends in the west.

An oft-overlooked constellation is **Camelopardalis (the Giraffe)**, located between Auriga and **Perseus (the Hero)**. It is home to a wonderful collection of stars known as **Kemble's Cascade**, an asterism made up of more than 20 faint stars that form a straight line that stretches roughly five Moon diameters. You can find it using 10 × 50 binoculars or a small telescope. Look for a misty star cluster **NGC 1502** at the end of the line.

This month **Draco (the Dragon)** is most prominent in the northeast as it slinks its way around **Ursa Minor (the Little Bear)**. High above is **Ursa Major (the Great Bear)**, with its asterism the **Big Dipper** upside down and spilling its contents.

This is a great time to look for the **Whirlpool Galaxy (Messier 51)**, which can be found above the Big Dipper's star **Alkaid**. Also nearby is the **Pinwheel Galaxy (Messier 101)**. You can find it forming a triangle below Alkaid and **Mizar**, the middle star in the Big Dipper's handle.

If you want more easy-to-spot galaxies, **Bode's Galaxy (Messier 81)** and the **Cigar Galaxy (Messier 82)** are in a great observing location this month.

Finally, to the east is **Boötes (the Herdsman)**, with its unmistakable and prominent star **Arcturus** at center stage.

May

May's Events

While the nights may be getting shorter, there are a few planets you can catch in the eastern sky before sunrise.

You can find Neptune and Saturn low in the east, with Mars rising above the horizon just before the Sun comes up. Unfortunately, Uranus and Mercury are too close to the Sun to be observed. Brilliant Venus graces the western sky after sunset, while almost-as-bright Jupiter continues to shine to its upper left, in Gemini.

The night sky is slowly beginning to change from the familiar winter and spring constellations to summer constellations, with Hercules climbing in the east and Cygnus and Lyra trailing behind him in the northeast.

Boötes is unmistakable, with Arcturus shining high in the southeast, while the winter constellations Auriga and Gemini drop low in the west with Jupiter.

Meanwhile, Ursa Major is overhead in the sky, making it an opportune time to take out a telescope and do some galaxy hunting in the Great Bear.

Calendar of Events

Day	Time (UTC)	Event
01	17:23	Full Moon
02	00:59	Aldebaran 6.4°N of Venus
04	02:20	Antares 0.5°N of the Moon
04	22:30	Moon at apogee: 405,800 km (252,152 mi.)
05		Eta Aquariids meteor shower peak
09	21:11	Last quarter of the Moon
16	20:01	New Moon
17	13:48	Moon at perigee: 358,100 km (222,513 mi.)
19	01:50	Venus 2.9°S of the Moon
20	12:39	Jupiter 3.1°S of the Moon
21	14:39	Beehive 0.8°S of the Moon
23	11:11	First quarter of the Moon
27	14:09	Spica 2.1°N of the Moon
31	08:32	Antares 0.4°N of the Moon
31	08:45	Full Moon

The Moon This Month

SUN	MON	TUES	WED	THURS	FRI	SAT
					1 Full Moon	2
3	4	5	6	7	8	9 Last Quarter
10	11	12	13	14	15	16 New Moon
17	18	19	20	21	22	23 1st Quarter
24	25	26	27	28	29	30
31 Full Moon						

On May 4, the bright Moon will shine near Antares after midnight. On May 13, a waning crescent Moon will join Saturn in the eastern morning sky together with much fainter Neptune nearby. On May 14, the pretty Moon will shine above and between Saturn and Mars above the eastern horizon before sunrise.

A few days later, on May 18, the Moon and Venus have a very close conjunction in the west after sunset. The Moon will be only 7 percent illuminated, and the pair will be less than 2 degrees apart.

Then, on May 19, Jupiter gets in on the Moon action when a waxing crescent that will be roughly 24 percent illuminated will join the planet in Gemini. And the next day, the Moon will shine between Jupiter and the Beehive Cluster.

This is a special month with not one, but two full Moons. The first full Moon falls on May 1, with the second — what is often referred to as a "blue Moon" — on May 31.

The last quarter is on May 9, the new Moon on May 16 and the first quarter on May 23.

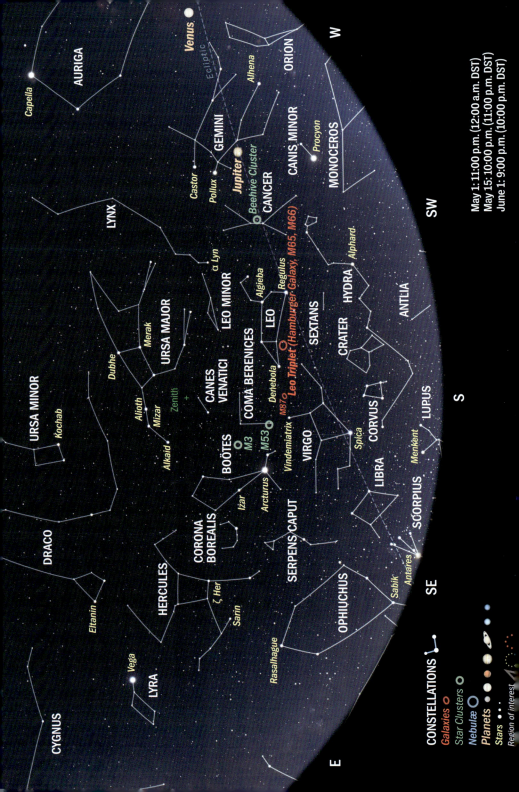

Leo Triplet (Messier 65, Messier 66 and NGC 3628).

Highlights in the Southern Sky

Boötes (the Herdsman) can be found high in the southeast now, along with **Corona Borealis (the Northern Crown)**. The latter is a small but unmistakable constellation with its U-shaped formation. It is known for its "blaze star," **T Coronae Borealis**, a recurrent nova that brightens roughly every 80 years. The last time it brightened was in 1946. This star may or may not have already "blazed" by the time of publication.

Near Boötes' star **Arcturus**, but in **Canes Venatici (the Hunting Dogs)**, you can find a nice globular cluster called **Messier 3**, which can be seen as a faint smudge in binoculars on a dark night. Nearby, in **Coma Berenices (Berenice's Hair)**, you can find another globular cluster, **Messier 53**, though it is a tad fainter.

In the sky between **Virgo (the Maiden)** and **Leo (the Lion)**, you can find the faint elliptical galaxy **Messier 87**. What's most notable about this enormous galaxy is that in 2019 an international group of astronomers from the Event Horizon Telescope revealed the first image of a black hole at its core.

Just below Leo, a modest telescope will reveal the **Leo Triplet**, a collection of three galaxies all within the same field of view. The group contains **Messier 65, Messier 66** and, perhaps the most famous galaxy of the trio, **NGC 3628** (also known as the **Hamburger Galaxy**). It got that nickname due to its layered shape.

Finally **Jupiter** is sinking low in the west along with **Gemini (the Twins)**, with its pair of stars **Castor** and **Pollux**.

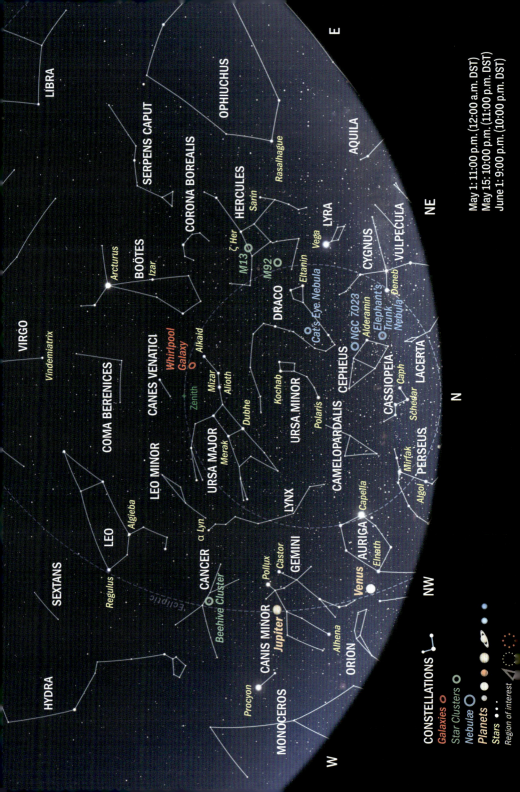

Cat's Eye Nebula (NGC 6543)

Highlights in the Northern Sky

In the northeast, the mighty **Hercules**, the fifth-largest constellation in the night sky, is rising. It is home to one of the most beautiful globular clusters in the northern sky, **Messier 13**. This cluster contains some 300,000 stars that are held together tightly by gravity. It lies roughly 22,200 light-years from Earth and has an estimated age of 11.65 billion years. Because it is one of the brightest globular clusters in the sky, it is easily visible from dark-sky sites or through binoculars. Another globular cluster in Hercules is **Messier 92**, though it is not quite as prolific as the "Great Globular Cluster in Hercules."

Cassiopeia (the Queen) is hugging the northern horizon but will be making the upward journey in the coming months, along with **Perseus (the Hero)**.

To the right of **Polaris**, but in the constellation **Draco (the Dragon)**, you can find the **Cat's Eye Nebula (NGC 6543)**, a planetary nebula located roughly 3,000 light-years away. Despite the name, planetary nebulae have nothing to do with planets. Instead, their nomenclature comes from some having a spherical appearance, similar to a planet.

Climbing the northeast, you can find the brilliant **Vega**, the brightest star in **Lyra (the Harp)**. As the weeks progress, nearby **Cygnus (the Swan)** will ascend in the east, though, at the moment, it is just above the northeastern horizon.

June

June's Events

We now welcome summer with its warmer but shorter nights. This time of year also welcomes the heart of the Milky Way to the southern sky.

Mars, Saturn, Uranus and Neptune can be found in the eastern sky in the early morning hours. On June 8 to 9, there is a nice conjunction of Jupiter and Venus in the west after sunset.

Around June 10, Mercury will be visible below the pair over the next several days.

On June 19, Venus and the Beehive Cluster have a close encounter in the west after sunset. This would make a nice binocular target for those looking for a close-up view.

The summer solstice falls on June 21.

Calendar of Events

Day	Time (UTC)	Event
01	04:32	Moon at apogee: 406,400 km (252,525 mi.)
07	16:15	Pollux 4.7°S of Venus
08	10:00	Last quarter of the Moon
09	20:11	Jupiter 1.6°N of Venus
13	13:15	Pleiades 0.9°S of the Moon
14	23:18	Moon at perigee: 357,200 km (221,954 mi.)
15	02:54	New Moon
15	19:59	Mercury 24.5°E of the Sun (greatest elongation east)
16	19:32	Mercury 2.6°S of the Moon
17	06:54	Jupiter 2.5°S of the Moon
17	20:21	Venus 0.3°S of the Moon
17	23:40	Beehive 0.6°S of the Moon
18	14:59	Pollux 6.4°S of Mercury
19	14:31	Regulus 0.3°N of the Moon
19	15:44	Beehive 0.4°N of Venus
21	08:25	Summer solstice
21	21:55	First quarter of the Moon
23	20:11	Spica 2.3°N of the Moon
25	11:54	Jupiter 3.8°N of Mercury
27	14:32	Antares 0.5°N of the Moon
28	07:11	Moon at apogee: 406,300 km (252,463 mi.)
28	18:30	Pleiades 4.4°S of Mars
29	23:57	Full Moon

The Moon This Month

SUN	MON	TUES	WED	THURS	FRI	SAT
	1	2	3	4	5	6
7	8 Last Quarter	9	10	11	12	13
14	15 New Moon	16	17	18	19	20
21 1st Quarter	22	23	24	25	26	27
28	29 Full Moon	30				

On June 10, a waning crescent Moon joins Saturn in the east in the early morning hours. On June 12, a thin waning crescent Moon shines above Mars in the east before sunrise. The next morning the Moon will sit near the Pleiades and Uranus. You'll need to live at a southerly latitude and have an unobstructed view of the horizon to catch the sight.

On June 16, a very thin waxing crescent Moon will shine near Venus, Jupiter and Mercury, which will all be lined up in the west after sunset for about a week. On June 17, there is a conjunction of the Moon and Venus near the Beehive Cluster.

The Moon's last quarter is on June 8, with the new Moon on June 15 (though it will be best viewed on the evening of June 14). The Moon is at first quarter on June 21, and the full Moon falls on June 29.

The first quarter Moon hangs with Mars.

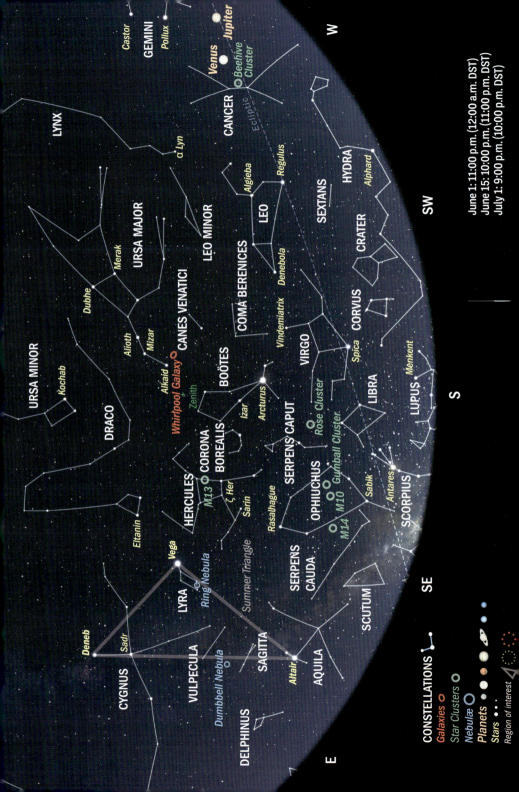

Dumbbell Nebula (Messier 27)

June 2026

Highlights in the Southern Sky

With true darkness falling so late around the June solstice, stargazers will need to stay up late to catch some of the most beautiful sights of the night sky. But it's worth it, with the richest part of the Milky Way dominating the south.

High in the southeast is **Ophiuchus (the Serpent Bearer)**. There are several beautiful globular clusters to be found here, including **Messier 10**, **Messier 12 (Gumball Cluster)** and **Messier 14** — all great sights in any binoculars, particularly under darker skies.

Nearby is **Serpens Caput (the Head of the Serpent)**, where you can find the impressive globular named the **Rose Cluster (Messier 5)**. But the jewel of globular clusters has to be **Messier 13**, which is well placed this month in **Hercules**.

The **Summer Triangle**, an asterism containing the brightest stars from three different constellations — **Deneb**, **Vega** and **Altair** — is now rising in the east.

Also to be found near Vega in **Lyra (the Harp)** is the small but impressive **Ring Nebula (Messier 57)**, a planetary nebula located 2,300 light-years from Earth. A moderately large telescope is needed to spot its ring-like shape and tiny central white dwarf star.

You can also find a favorite of many astrophotographers, the **Dumbbell Nebula (Messier 27)** in **Vulpecula (the Little Fox)**, a small constellation that lies between **Cygnus (the Swan)** and **Aquila (the Eagle)**.

June 81

Caroline's Rose (NGC 7789)

Highlights in the Northern Sky

Cygnus (the Swan) can be found high in the northeast, with its brightest star **Deneb** marking the bird's tail. Near to Deneb is the **North America Nebula (NGC 7000)**, a rich and fairly bright emission nebula that glows due to the ionization of interstellar gas caused by stellar radiation. It's located roughly 1,700 light-years from Earth and is 100 light-years across. **Messier 39**, a bright open cluster that lies only 824 light-years from Earth, sits to Deneb's lower left. It is best viewed through binoculars or a small telescope.

Turning our gaze northeast, we come to **Cepheus (the King)**. This is where you can find **Mu (μ) Cephei (the Garnet Star)**, a red supergiant. The star has a radius that is anywhere from 1,200 to 1,600 times that of the Sun and is one of the largest stars visible to the unaided eye.

The Garnet Star is on the edge of the **Elephant's Trunk Nebula (IC 1396)**, a dense region of dust and gas, with a long lobe that has given it its name.

While **Cassiopeia (the Queen)** is still fairly low in the north, its familiar W shape is beginning to rise. **Caroline's Rose (NGC 7789)**, a nice open cluster, can be found in Cassiopeia. The cluster is estimated to be roughly 1.7 billion years old, making it one of the oldest open clusters in the northern celestial hemisphere. It is best seen with binoculars or a small telescope.

July

July's Events

The Milky Way is now magnificently stretching across the sky, from Cassiopeia, through Cygnus and Aquila, to Sagittarius and Scorpius. Medium-sized binoculars will reveal just how rich with stars the region is.

If you have binoculars, summer is the best time to use them, with the center of the Milky Way prominent in the south. There are numerous clusters, both globular and open, as well as the unmistakable Sagittarius Star Cloud.

Saturn and Neptune are now rising in the east around midnight, and Mars and Uranus are both hosted by Taurus in the east before sunrise. Jupiter is now lost in the Sun's glare in the evening, but Venus is still shining brightly in the west after sunset.

On July 4, there is a very tight conjunction of Mars and Uranus in the morning sky, but you'll need a clear view of the eastern horizon. The pair will also sit midway between the Hyades and Pleiades clusters, so the planets should be fairly easy to spot with binoculars.

Earth is at aphelion, or the farthest from the Sun in its annual orbit, on July 6.

Calendar of Events

Day	Time (UTC)	Event
06	14:59	Earth at aphelion: 1.0166 astronomical units (au)
07	19:29	Last quarter of the Moon
09	14:36	Regulus 1°N of Venus
10	22:54	Pleiades 1.1°S of the Moon
12	13:17	Aldebaran 5.3°N of Mars
13	07:50	Moon at perigee: 359,100 km (223,134 mi.)
14	09:43	New Moon
17	00:07	Regulus 0.5°N of the Moon
17	16:31	Venus 2.1°N of the Moon
21	03:21	Spica 2.5°N of the Moon
21	11:06	First quarter of the Moon
24	21:00	Antares 0.6°N of the Moon
25	16:45	Moon at apogee: 405,500 km (251,966 mi.)
29	14:36	Full Moon

The Moon This Month

SUN	MON	TUES	WED	THURS	FRI	SAT
			1	2	3	4
5	6	7 Last Quarter	8	9	10	11
12	13	14 New Moon	15	16	17	18
19	20	21 1st Quarter	22	23	24	25
26	27	28	29 Full Moon	30	31	

On July 7, the last quarter Moon will sit above and between Saturn and Neptune in the eastern sky.

When the waning crescent Moon meets up with Mars before sunrise on July 11, they'll be near the Pleiades as well as the Hyades and Uranus.

On July 16 and 17, a waning crescent Moon pairs with Venus in the western sky after dusk.

The new Moon is on July 14, first quarter falls on July 21, and the full Moon occurs on July 29.

The waning crescent Moon with Venus

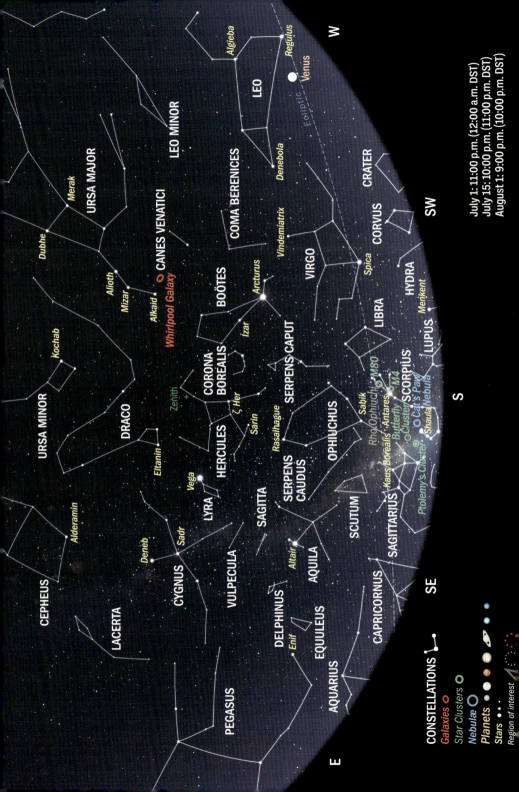

Cat's Paw Nebula (NGC 6334)

Highlights in the Southern Sky

As the nights are now shorter, it means that observers will need to stay up later to take in some of summer's most spectacular sights. But it's certainly worth it.

Looking directly south is the richest region of the Milky Way. **Scorpius (the Scorpion)** can't be missed due to its brightest star, the "heart" of the constellation, **Antares**. This is a red supergiant, just like Betelgeuse in Orion. It is one of the largest known stars, roughly 700 times larger in diameter than our Sun.

Nearby, is the bright globular cluster **Messier 4**, a collection of tens of thousands of stars. It is also one of the closest globular clusters to Earth at roughly 7,200 light-years.

Above Antares you can find the **Rho Ophiuchi cloud complex**, which contains both bright and dark nebulae. It is located roughly 460 light-years away and is the nearest star-forming region to us. You can find another fairly bright globular cluster, **Messier 80**, beside the cloud complex.

Near the "tail" of the scorpion, close to **Sagittarius (the Archer)**, are two globular clusters. The brightest of the two is **Ptolemy's Cluster (Messier 7)**. It is bright enough to be seen with the naked eye, particularly from dark-sky locations. The second is the slightly dimmer **Butterfly Cluster (Messier 6)**, found nearby.

Off Scorpius' tail star **Shaula** is the **Cat's Paw Nebula (NGC 6334)**, an emission nebula that is roughly 50 light-years across and lies 5,500 light-years away.

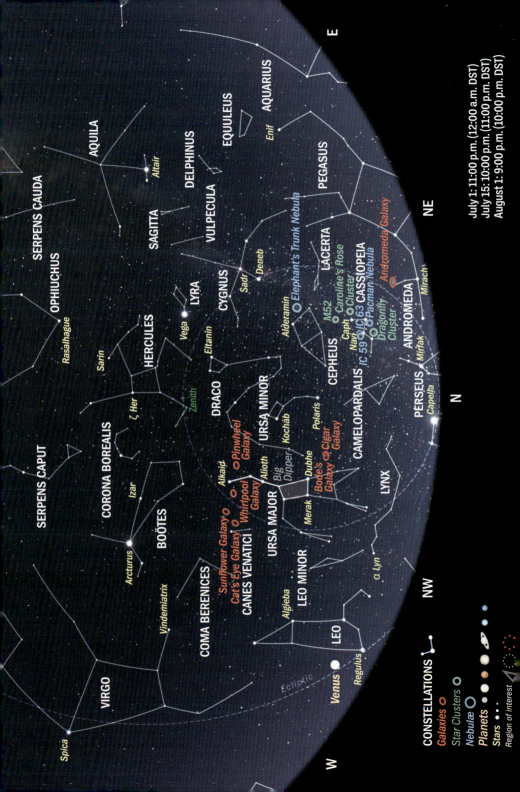

Sunflower Galaxy (Messier 63)

Highlights in the Northern Sky

Cassiopeia (the Queen) is now rising in the northeast, with its familiar "W" shape. Two beautiful nebulae lie here, right off the star **Navi**, which can be found in the middle of the "W." **IC 59** and **IC 63**, known as the **Ghost of Cassiopeia**, are two emission and reflection nebulae that are best seen through large telescopes.

Also nearby is the brighter **Pacman Nebula (NGC 281)**, a large emission nebula and star-forming region that lies roughly 9,200 light-years from Earth.

Of the many open clusters in Cassiopeia, three bright ones to check out are the **Dragonfly Cluster (NGC 457)**, **Messier 52** and **Caroline's Rose (NGC 7789)**.

Over in the northwest is **Ursa Major (the Great Bear)**. July is a great time to do some galaxy hunting in this constellation. **Bode's Galaxy (Messier 81)** and the **Cigar Galaxy (Messier 82)** are well placed for viewing, as is the **Pinwheel Galaxy (Messier 101)**.

Between **Alkaid**, the last star in the Big Dipper asterism, and **Canes Venatici (the Hunting Dogs)**, you can find the **Whirlpool Galaxy (Messier 51)**. The **Sunflower Galaxy (Messier 63)** and the **Cat's Eye Galaxy (Messier 94)** also lie nearby.

August

August's Events

Summer may be winding down, but there are still plenty of nights to enjoy the season's cosmic sights.

Sagittarius is still well placed for viewing and Ophiuchus is high in the southwest. This month the Summer Triangle asterism dominates the sky.

In the east you can find Saturn and Neptune in Pisces in late evening. Mars is in Gemini and climbs the eastern sky in the early morning hours, led by Uranus in Taurus. On August 15 there is a very tight conjunction of Jupiter and Mercury; however, the pair will be extremely low on the horizon in the east just before sunrise.

There is a total solar eclipse on August 12 over the North Pole, Greenland, Iceland and Spain. In Canada and the northeastern U.S., it will be a partial solar eclipse. On August 28, there is a partial lunar eclipse that will be visible across North America and South America.

The highlight of the month is the most anticipated show of the year, the Perseids meteor shower — a favorite of even the casual stargazer. The shower rarely disappoints, with dozens of meteors seen in an hour (under ideal conditions), often with long trains. This year, the Perseids peak on the night of August 12 to 13. Fortunately, it's good news for skywatchers: it will be a new Moon, meaning you can enjoy what is likely to be a wonderful show.

Calendar of Events

Day	Time (UTC)	Event
02	07:59	Mercury 19.5°W of the Sun (greatest elongation west)
06	02:21	Last quarter of the Moon
07	06:23	Pleiades 1.2°S of the Moon
09	05:31	Mars 4.4°S of the Moon
10	11:18	Moon at perigee: 363,300 km (225,744 mi.)
12	17:37	New Moon
12	17:47	Total solar eclipse (partial solar eclipse observable from North America)
12		Perseids meteor shower peak
15	05:59	Venus 45.9°E of the Sun (greatest elongation east)
16	08:46	Venus 2.2°N of the Moon
17	11:49	Spica 2.6°N of the Moon
20	02:46	First quarter of the Moon
21	04:17	Antares 0.7°N of the Moon
22	08:20	Moon at apogee: 404,600 km (251,408 mi.)
28	04:14	Deep partial lunar eclipse
28	04:18	Full Moon

The Moon This Month

August 2026

SUN	MON	TUES	WED	THURS	FRI	SAT
						1
2	3	4	5	6 Last Quarter	7	8
9	10	11	12 New Moon	13	14	15
16	17	18	19	20 1st Quarter	21	22
23	24	25	26	27	28 Full Moon	29
30	31					

Overnight on August 2 and 3, a waning gibbous Moon meets up with Saturn. In the early hours of August 7, it swings through the Pleiades, and two days later, on August 9, the waning crescent Moon and Mars meet up in the east. This is best seen before sunrise.

There is a conjunction between the old crescent Moon and Mercury in the east before sunrise on August 11. The young crescent Moon will join Venus on August 15, though it will be very low on the western horizon after sunset.

It will shine higher, to Venus' left, the next evening. At the end of the month, on August 30, the bright waning gibbous Moon will shine near Saturn and Neptune in the east after dusk. Early risers can catch them in southwest before sunrise on August 31.

The last quarter of the Moon falls on August 6, the new Moon is on August 12, the first quarter is on August 20 and the full Moon rises on August 28, in time for a partial lunar eclipse.

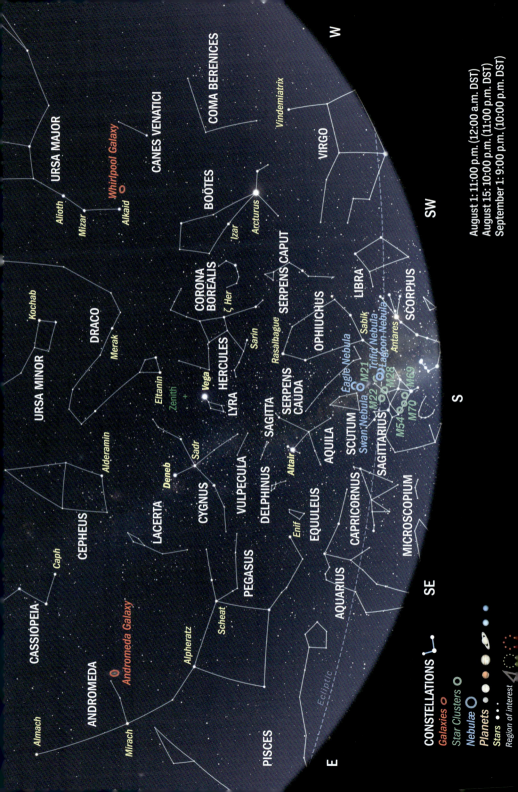

Lagoon Nebula (Messier 8) and, above it, the Trifid Nebula (Messier 20)

Highlights in the Southern Sky

Beginning in the south is **Sagittarius (the Archer)**. The easiest way to spot the constellation is to look for its teapot-shaped asterism. There are three nice globular clusters here: **Messier 54**, **Messier 69** and **Messier 70**, all found near the bottom of the teapot.

Near the top of the teapot are two impressive globular clusters, the **Great Sagittarius Cluster (Messier 22)**, which is quite bright, and **Messier 28**.

Just hugging the boundary of Sagittarius, **Scutum (the Shield)** and **Serpens Cauda (the Tail of the Serpent)** is the **Swan Nebula (Messier 17)**. Also known as the **Omega Nebula**, this star-forming region is best seen through binoculars or a small telescope.

Sticking with Sagittarius, we can find the **Lagoon Nebula (Messier 8)**, a bright star-forming region that lies 4,100 light-years from Earth. At its center is a cluster of stars, **NGC 6523**. The nebula is bright enough to be seen through decent binoculars but is better seen with a small telescope.

Just above the Lagoon Nebula is the **Trifid Nebula (Messier 20)**, which is made up of a reflection nebula, an emission nebula, a dark nebula and an open cluster. And just above the Trifid is a nice, bright open cluster, **Messier 21**.

Moving up to Scutum, you can find the **Eagle Nebula (Messier 16)**, made famous by the Hubble Space Telescope in 1995 when it captured the nebula's soaring tendrils of cosmic dust and gas. These features, which sit at the heart of the nebula, were given the nickname "the Pillars of Creation."

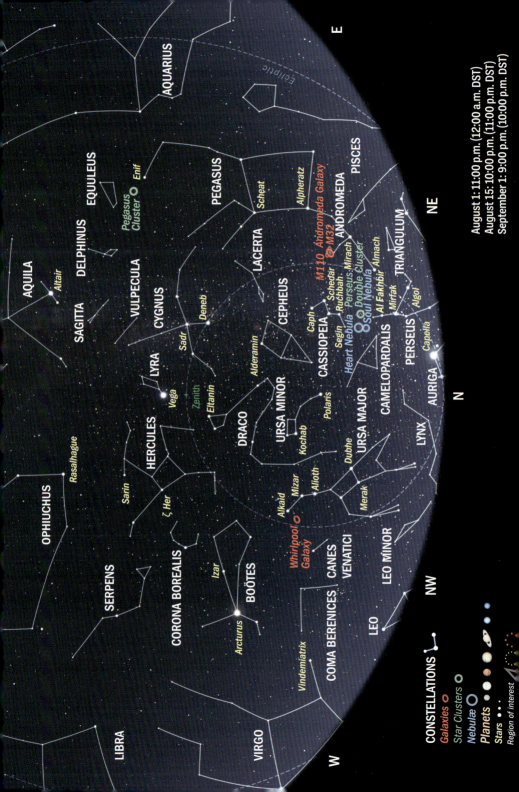

Highlights in the Northern Sky

Perseus (the Hero) can be found low in the northeast. The most notable feature in this constellation is the **Perseus Double Cluster (NGC 869 and NGC 884)**, two open star clusters that can be seen in the same field of view through binoculars or a small telescope. The pair lie roughly 7,000 light-years from Earth and are separated by only a few hundred light-years. Follow **Mirfak**, the brightest star in Perseus, up toward **Al Fakhbir** to lead you to the two open clusters.

Also nearby, you can find the **Soul Nebula (IC 1848)** and the **Heart Nebula (IC 1805)**, both of which lie in **Cassiopeia (the Queen)**. Those appear best in large telescopes and long-exposure photographs.

Following Cassiopeia's star **Schedar** toward **Mirach** in **Andromeda (the Princess)**, you will find the **Andromeda Galaxy (Messier 31)** along with two smaller galaxies, **Messier 32** and **Messier 110**. The Andromeda Galaxy, which lies roughly 2.54 million light-years away and is on a collision course with our own galaxy, is visible to the naked eye from dark-sky locations and looks wonderful in binoculars.

Lastly, in the northeast is the mighty **Pegasus (the Flying Horse)**. Beyond his nose star **Enif** you can find the **Pegasus Cluster (Messier 15)**, a fairly bright globular cluster that is easily seen in quality binoculars and backyard telescopes.

From left to right, the Soul Nebula (IC 1848) and Heart Nebula (IC 1805)

September

September's Events

Cooler nights are now upon us, but that also means it's darker earlier and the sky is steadier in terms of atmospheric turbulence from daytime heating. It's a give and take with astronomy.

The summer constellations are sinking low in the southwest, but we now get to see Perseus, Cassiopeia and Pegasus in the east. And for observers at dark-sky locations, the Andromeda Galaxy will be easy to spot with the naked eye.

Venus is still shining brightly very low in the west after sunset. The "evening star" will be at its brightest on September 22. Saturn is now moving retrograde west into Cetus, leaving Neptune in next-door Pisces. Both planets rise in the east an hour or so after sunset. Uranus is in Taurus and rises in late evening. And if you're up early you can find Mars in Gemini, near Pollux. Finally, Jupiter rises in the east in Cancer a few hours before sunrise.

The autumnal equinox falls on September 22.

Calendar of Events

Day	Time (UTC)	Event
01	13:27	Spica 1.4°S of Venus
03	12:03	Pleiades 1.2°S of the Moon
04	07:51	Last quarter of the Moon
06	18:24	Mars 3°S of the Moon
06	20:26	Moon at perigee: 368,300 km (228,851 mi.)
08	04:43	Beehive 0.5°S of the Moon
08	18:13	Jupiter 0.8°S of the Moon
11	03:27	New Moon
13	20:53	Spica 2.6°N of the Moon
14	11:10	Venus 0.6°S of the Moon
17	01:42	Pollux 5.9°S of Mars
17	12:18	Antares 0.6°N of the Moon
18	20:44	First quarter of the Moon
19	03:00	Moon at apogee: 404,200 km (251,158 mi.)
23	00:06	Autumnal equinox
26	00:13	Neptune at opposition
26	01:49	Spica 0.9°N of Mercury
26	16:49	Full Moon
30	17:39	Pleiades 1.1°S of the Moon

The Moon This Month

SUN	MON	TUES	WED	THURS	FRI	SAT
		1	2	3	4 Last Quarter	5
6	7	8	9	10	11 New Moon	12
13	14	15	16	17	18 1st Quarter	19
20	21	22	23	24	25	26 Full Moon
27	28	29	30			

This month the Moon has a busy social calendar as it meets up with many targets in the night sky. Starting late on September 2, the waning gibbous Moon hangs out with the Pleiades, and on September 6 and 7, the waning crescent Moon joins Mars in the eastern sky in the early morning hours. A day later, on September 8, there is a conjunction of a thin crescent Moon and Jupiter near the Beehive Cluster in the east a few hours before sunrise. On September 13 and 14, the crescent Moon joins Venus low in the west after sunset. On September 26, there is a conjunction of an almost full Moon and Saturn, with Neptune nearby. And lastly, on September 30, the waning gibbous Moon meets with the Pleiades for observers in the Pacific region.

The Moon is at last quarter on September 4, with the new Moon falling on September 11 (though best viewed on September 10). First quarter is on September 18, and the full Moon takes place on September 26.

The Pleiades (Messier 45) will be visited by the Moon twice in September

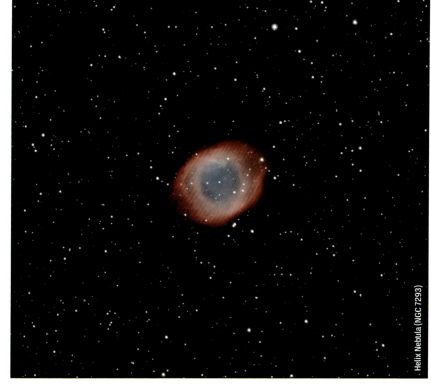

Helix Nebula (NGC 7293)

Highlights in the Southern Sky

In the southeast lies **Aquarius (the Water Bearer)**. Near the bottom of the constellation you can find the **Helix Nebula (NGC 7293)**, sometimes referred to as the "eye of God." This is a fairly bright and large planetary nebula that lies roughly 650 light-years from Earth. It is one of the closest and brightest planetary nebulae.

To the north of Aquarius is **Messier 2**, a reasonably bright globular cluster that contains over 150,000 stars.

Next door, in **Capricornus (the Sea Goat)**, is the **Jellyfish Cluster (Messier 30)**, another globular cluster that is easily seen in good binoculars.

High in the southeast, you will spot the tiny but unmistakable **Delphinus (the Dolphin)**. This constellation looks like a tiny kite. Another globular cluster, **NGC 6934**, sits beyond the kite's string.

To the lower right of Delphinus, **Aquila (the Eagle)** is marked by its brightest star, **Altair**. This is an interesting star. It rotates extremely quickly, once every 8.9 hours. This causes its equatorial regions to bulge out, which creates an effect called "gravity darkening," where the equatorial zone is cooler and less luminous than the poles.

Below Aquila is **Scutum (the Shield)**, which contains the prominent **Wild Duck Cluster (Messier 11)** and **Messier 26**, both open clusters.

Ophiuchus (the Serpent Bearer) in the southwestern sky has several globular clusters, including **Messier 10** and the **Gumball Cluster (Messier 12)**, which can be found "inside" the constellation, as well as **Messier 14**, which lies closer to the bright star **Cebalrai**. Finally, there is **Messier 107**, which can be found below the star **Zeta Ophiuchi**.

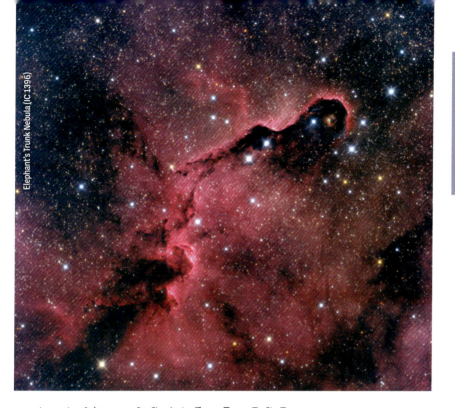

Elephant's Trunk Nebula (IC 1396)

September 2026

Highlights in the Northern Sky

Ursa Major (the Great Bear) now appears to be walking along the northwest to the northern horizon.

In the west you can find **Boötes (the Herdsman)** sinking near the horizon. Above its star **Arcturus** is **Corona Borealis (the Northern Crown)**, with **Hercules** above, where you can find the beautiful globular cluster **Messier 13**.

But there's a lot more to see in the northeast and near the zenith.

Perseus (the Hero) is rising higher in the northeast. It contains a fairly bright globular cluster **Messier 34**, which can be found above **Algol**. The star **Miram** points to the popular **Double Cluster (NGC 869 and NGC 884)**.

In **Andromeda (the Princess)** you can see a large, bright but scattered open cluster **NGC 752** and, of course, the breathtaking **Andromeda Galaxy (Messier 31)**.

In **Triangulum (the Triangle)** is the large but very faint **Triangulum Galaxy (Messier 33)**, which lies roughly 2.7 million light-years away.

Meanwhile, **Cepheus (the King)** is approaching the zenith. You can find the **Elephant's Trunk Nebula (IC 1396)** along the base and near the middle of its inverted, house-shaped constellation. In the same location is **Mu (μ) Cephei (the Garnet Star)**, a red supergiant.

September 101

October

October's Events

This month the Summer Triangle asterism will catch your eye in the west, while Pegasus dominates the southern sky. Cassiopeia and Perseus are now well placed for viewing in the northeast. This is also a great time of year to see the Andromeda Galaxy, as it is almost at the zenith.

Saturn and Neptune are now observable all night long, as is Uranus, though it rises in mid-evening. Jupiter and Mars are in the east in the early morning, with Mars spectacularly crossing the Beehive Cluster on October 11. Unfortunately, Venus is now lost in the Sun's glare.

Finally, the Orionids meteor shower peaks on the night of October 21 to 22, though the bright waxing gibbous Moon will interfere with all but the brightest meteors.

Calendar of Events

Day	Time (UTC)	Event
01	20:41	Moon at perigee: 369,300 km (229,472 mi.)
03	13:25	Last quarter of the Moon
04	11:52	Saturn at opposition
05	05:30	Mars 1.2°S of the Moon
05	11:06	Beehive 0.4°S of the Moon
06	10:18	Jupiter 0.2°S of the Moon
07	02:57	Regulus 0.6°N of the Moon
07	03:59	Venus 5°N of Mercury
10	15:50	New Moon
10	15:55	Beehive 0.4°S of Mars
12	02:30	Venus 3.3°S of the Moon
12	09:59	Mercury 25.2°E of the Sun (greatest elongation east)
12	20:08	Mercury 2.3°N of the Moon
14	20:25	Antares 0.5°N of the Moon
16	22:56	Moon at apogee: 404,600 km (251,407 mi.)
18	16:13	First quarter of the Moon
21		Orionids meteor shower peak
26	04:12	Full Moon
28	01:11	Pleiades 1°S of the Moon
28	18:01	Moon at perigee: 364,400 km (226,428 mi.)

The Moon This Month

SUN	MON	TUES	WED	THURS	FRI	SAT
				1	2	3 Last Quarter
4	5	6	7	8	9	10 New Moon
11	12	13	14	15	16	17
18 1st Quarter	19	20	21	22	23	24
25	26 Full Moon	27	28	29	30	31

October 2026

On October 5, there is a close conjunction of a waning crescent Moon and Mars, and as an added bonus, the pair will be close to the Beehive Cluster. You can find them in the east a couple of hours before sunrise.

The real treat comes the next day, on October 6, when there is a conjunction of the Moon and Jupiter a few hours before sunrise. And in some parts of Canada and the U.S. the Moon will pass in front of the giant planet in a lunar occultation of Jupiter.

The young crescent Moon will shine near Mercury above the western horizon after sunset on October 12. An almost full Moon meets up with Saturn on October 23 and 24, and on October 27, the Moon will cross through the Pleiades in the east.

Last quarter falls on October 3, with the new Moon taking place on October 10. First quarter occurs on October 18, and the full Moon falls on October 26.

The Moon passes in front of Jupiter this month

October 103

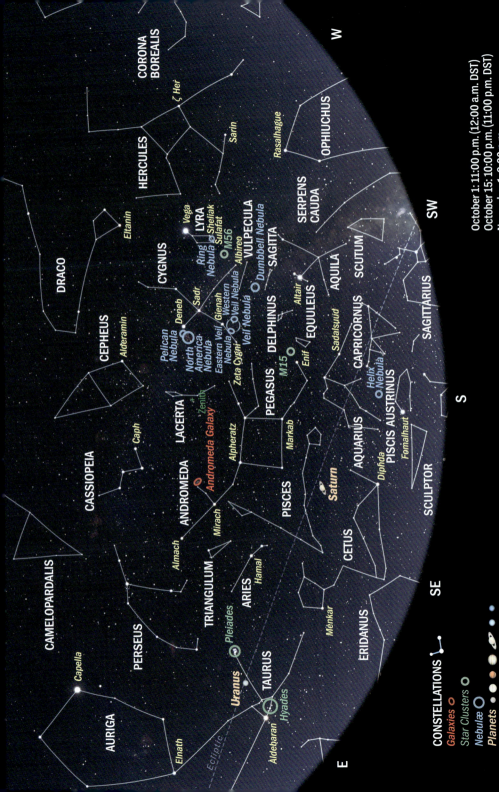

Pelican Nebula (IC 5070)

Highlights in the Southern Sky

Cygnus (the Swan) is high in the west near the zenith. It's a perfect time to check out its treasures, particularly if you're using binoculars or a telescope. The constellation contains several remarkable nebulae and is also a region rich with stars.

Off **Deneb**, the brightest star in Cygnus, is the **North America Nebula (NGC 7000)**, which is best seen through large telescopes. Also nearby is the **Pelican Nebula (IC 5070)**, an emission nebula about 30 light-years across and 1,800 light-years from Earth.

In between two of Cygnus' "wing" stars, **Gienah** and **Zeta Cygni**, is the wispy **Veil Nebula**. This is actually broken into the **Eastern Veil Nebula (NGC 6992)** and **Western Veil Nebula (NGC 6960)**. The heated ionized gas is a result of an ancient supernova explosion that is now five times the diameter of the full Moon!

Off **Albireo**, the second-brightest star in Cygnus, is the globular cluster **Messier 56**, which is actually found in **Lyra (the Harp)**.

And sticking with Lyra, you can find the **Ring Nebula (Messier 57)** between the stars **Sulafat** and **Sheliak**. This is best seen though a medium-sized telescope.

Moving to **Vulpecula (the Little Fox)**, you can find the **Dumbbell Nebula (Messier 27)**. It is the second-brightest planetary nebula in the night sky, just behind the **Helix Nebula (NGC 7293)** — which is in the southern sky this month — and is easy to spot using binoculars.

Pegasus (the Winged Horse) is quite prominent in the southern sky. Off **Enif**, which marks the steed's nose, you can find **Messier 15**, a bright globular cluster that's sometimes referred to as the **Pegasus Cluster.**

October 2026

California Nebula (NGC 1499)

October 2026

Highlights in the Northern Sky

You may have noticed a particularly bright star rising in the northeast. That would be **Capella**, the brightest star in **Auriga (the Charioteer)**. This constellation will rise higher in the coming months.

Perseus (the Hero) is in a great position and offers a wonderful opportunity to catch the **California Nebula (NGC 1499)**. Though the nebula is considered visually bright, it has a low surface brightness that makes it challenging to observe.

You can also find a nice, bright open cluster **Messier 34** in Perseus, near the star **Algol**, which dips noticeably in brightness every 2.87 days.

Cassiopeia (the Queen) is now looking more like an "M" than a "W" high in the north. This also means the **Andromeda Galaxy (Messier 31)** is very well placed for viewing this month. Binoculars will reveal the giant spiral galaxy as a smudge, but small telescopes will reveal it in more detail.

Nearby is the **Triangulum Galaxy (Messier 33)**, the second-closest spiral galaxy to our own Milky Way, found in **Triangulum (the Triangle)**. Though it is large due to its closeness to us, it has a low surface brightness, which makes it a challenging target to observe.

To the west is **Hercules**, where you can find the incredible globular cluster **Messier 13**, one of the brightest and most well-known globular clusters in the northern sky. It is 22,200 light-years from Earth, contains several hundred thousand stars and is estimated to be more than 11 billion years old.

Another globular cluster, **Messier 92**, also sits nearby, though it is slightly dimmer than Messier 13.

November

November's Events

In November, Saturn and Neptune continue to be visible all night, as is Uranus in Taurus. Jupiter and Mars, in Leo in the east after midnight, meet in a close conjunction on November 14 and 15.

There are three meteor showers this month. The Southern Taurids peak on the night of November 5 to 6, and the Northern Taurids peak on the night of November 12 to 13 — both showers without any moonlight to wash them out. The Leonids meteor shower, which peaks on the night of November 17 to 18, will be best viewed after the Moon sets at midnight.

Calendar of Events

Day	Time (UTC)	Event
01	16:34	Beehive 0.2°S of the Moon
01	20:28	Last quarter of the Moon
02	14:23	Mars 1.1°N of the Moon
02	23:11	Jupiter 0.5°N of the Moon
03	08:40	Regulus 0.8°N of the Moon
05		Southern Taurids meteor shower peak
07	11:31	Venus 1.1°N of the Moon
09	07:02	New Moon
10	13:31	Spica 0.9°S of Venus
11	03:58	Antares 0.3°N of the Moon
12		Northern Taurids meteor shower peak
13	17:50	Moon at apogee: 405,600 km (252,028 mi.) SUPERMOON
16	04:22	Jupiter 1.2°N of Mars
17	11:48	First quarter of the Moon
18		Leonids meteor shower peak
20	22:59	Mercury 19.6°W of the Sun (greatest elongation west)
24	14:53	Full Moon
25	07:47	Regulus 1.7°N of Mars
25	20:58	Moon at perigee: 359,300 km (223,259 mi.)
25	22:47	Uranus at opposition
28	23:21	Beehive 0.1°N of the Moon
30	09:18	Jupiter 1.2°N of the Moon
30	14:35	Regulus 1.1°N of the Moon

The Moon This Month

SUN	MON	TUES	WED	THURS	FRI	SAT
1 Last Quarter	2	3	4	5	6	7
8	9 New Moon	10	11	12	13	14
15	16	17 1st Quarter	18	19	20	21
22	23	24 Full Moon	25	26	27	28
29	30					

The waning gibbous Moon shines near the Beehive Cluster in the early hours of November 1, and the last quarter Moon joins Mars and Jupiter on November 2. The next night, on November 3, Mars, Jupiter and the Moon all align in the east. The old crescent Moon will pose close to brilliant Venus in the east before sunrise on November 7.

Later this month, the bright gibbous Moon meets with Saturn in the evening on November 20. Then, on November 23 and 24, the almost-full Moon meets up with both the Pleiades and Hyades. The waning gibbous Moon reunites with the Beehive Cluster on November 28. Finally, starting late on November 29, there is a conjunction of the Moon and Jupiter in the east, with Mars nearby. Early risers on November 30 can see them before sunrise.

The Moon will be at last quarter on November 1, with the new Moon falling on November 9. First quarter occurs on November 17 and the full Moon on November 24.

The Moon greets the Beehive Cluster (Messier 44) twice in November

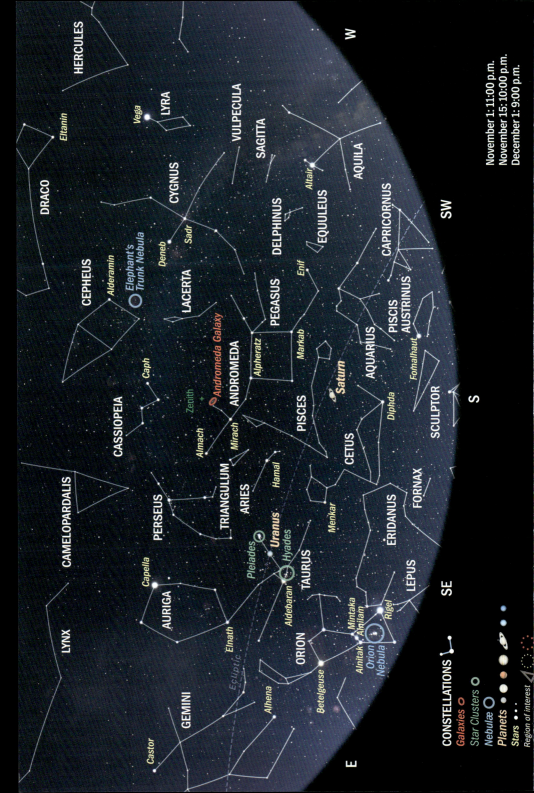

November 2026

The constellation Orion rises in the southeast with its many treasures

Highlights in the Southern Sky

We have **Orion (the Hunter)**, one of the northern sky's most unmistakable constellations, rising to the southeast. Its red supergiant star **Betelgeuse** is quite visible, even to the unaided eye, and its brightest star **Rigel** is also prominent. The three "belt stars" of the hunter — **Alnitak, Alnilam** and **Mintaka** — form the most distinctive part of the constellation.

And, of course, you can't leave out the **Orion Nebula (Messier 42)** in the "sword" of Orion. This incredible stellar nursery is visible to the unaided eye, even in moderately dark skies, and is one of the most popular targets for astrophotographers, who can capture its soft red and pink colors. Even medium-sized binoculars will reveal its structure.

However, two of the most beautiful objects to see are the **Pleiades (Messier 45)** and **Hyades (Melotte 25)**. The clusters are best seen with binoculars, and they won't disappoint — particularly Hyades. One of the most interesting features of Hyades is **Aldebaran**, a red giant that seemingly lies within the cluster but is not part of the collection of stars. In fact, it is roughly 80 light-years closer to Earth than the open cluster.

If you have large binoculars or a small telescope, this month you can find **Uranus** nearby, in between the Pleiades and Hyades. Over in the western sky, **Neptune** and **Saturn** are near one another in **Pisces (the Fishes)** and **Cetus (the Whale)**, respectively.

Near the zenith you can find **Andromeda (the Princess)** extending into **Pegasus (the Winged Horse)**.

November 111

Salt and Pepper Cluster (Messier 37)

Highlights in the Northern Sky

The **Little Dipper** asterism in **Ursa Minor (the Little Bear)** is now dangling downward in the northern sky, while **Ursa Major (the Great Bear)** is beginning to climb higher again in the northeast.

The **Cat's Eye Nebula (NGC 6543)** in **Draco (the Dragon)** was one of the first planetary nebulae to be discovered. These types of nebulae occur when a smaller, Sun-like star runs out of fuel, collapses and then expels its outer layers into space, leaving its hot core as a white dwarf star. Our Sun will one day become a planetary nebula — in about 5 billion years!

High above is **Cassiopeia (the Queen)**, with its rich treasure trove of celestial delights. Near the star **Caph** is **Caroline's Rose (NGC 7789)**, a beautiful open cluster. There are a few lovely open clusters in this prominent constellation, including **NGC 663** and the **Owl Cluster (NGC 457)**.

If you're into clusters, then you can't miss the **Double Cluster (NGC 869 and NGC 884)**, found in **Perseus (the Hero)**. The pair is very well placed in the sky, close to the zenith.

There's another bright open cluster in Perseus, **Messier 34**. This cluster lies roughly 1,500 light-years from Earth and is easy to spot using medium-sized binoculars.

Auriga (the Charioteer) is well situated this month. Its brightest star **Capella** dominates the northeast. Within the oval constellation are a few bright open clusters, the **Pinwheel Cluster (Messier 36)** and the **Starfish Cluster (Messier 38)**. Just outside of it is the **Salt and Pepper Cluster (Messier 37)**. All of these are clearly visible in binoculars.

Another bright open cluster is the **Shoe Buckle Cluster (Messier 35)**, found nearby in **Gemini (the Twins)**.

November 2026

December

December's Events

We have almost come full circle — around the Sun that is.

The winter constellations are taking hold. Orion is the most striking in the southeastern sky. Gemini and Taurus are also quite prominent, with Auriga high above them. Cepheus can now be found in the northwest, along with Cassiopeia.

Saturn and Neptune are beginning to sink in the southwest, along with the mighty Pegasus and Andromeda. Uranus is observable all night long in Taurus. Jupiter is rising in the east along with Mars a few hours after sunset. Meanwhile, Venus becomes the "morning star" in Libra, as it is visible a few hours before sunrise. Mercury will shine over the southeastern horizon before sunrise in early December.

The usually impressive Geminids meteor shower peaks on the night of December 13 to 14. Luckily, a waxing crescent Moon shouldn't interfere with the show.

The Ursids meteor shower peaks on December 22 to 23, but an almost-full Moon will interfere with all but the brightest meteors.

The winter solstice falls on December 21.

Calendar of Events

Day	Time (UTC)	Event
01	06:09	Last quarter of the Moon
04	18:36	Spica 2.7°N of the Moon
09	00:52	New Moon
11	06:46	Moon at apogee: 406,400 km (252,525 mi.)
12	17:41	Regulus 2.8°N of Jupiter
13		Geminids meteor shower peak
17	05:43	First quarter of the Moon
21	20:50	Winter solstice
21	22:37	Pleiades 1°S of the Moon
22		Ursids meteor shower peak
24	01:28	Full Moon
24	08:29	Moon at perigee: 356,700 km (221,643 mi.)
26	08:52	Beehive 0.3°N of the Moon
27	17:32	Jupiter 1.5°N of the Moon
27	22:44	Regulus 1.4°N of the Moon
30	18:59	Last quarter of the Moon

The Moon This Month

SUN	MON	TUES	WED	THURS	FRI	SAT
		1 Last Quarter	2	3	4	5
6	7	8	9 New Moon	10	11	12
13	14	15	16	17 1st Quarter	18	19
20	21	22	23	24 Full Moon	25	26
27	28	29	30 Last Quarter	31		

Before sunrise on December 1, the crescent Moon will shine with Jupiter and Mars. On December 4 and 5, a pretty waning crescent Moon pairs with Venus in the eastern sky before sunrise. Two days later, on December 7, a very thin crescent Moon — illuminated roughly 2 percent — shines near with Mercury above the southeastern horizon before sunrise. This will be a challenging conjunction to observe. A waxing gibbous Moon passes close by Saturn and Neptune on December 17 and then will pass through the Pleiades again a few days later, on December 21. On December 26, the Moon, Jupiter, the star Regulus and Mars will line up in the eastern sky after dusk. On December 27, the Moon will shine between Mars, Regulus and Jupiter. This will make a nice photo op and view through binoculars in the late evening.

The Moon is at last quarter on December 1 and the new Moon is on December 9 (but best seen on the night of December 8). First quarter occurs on December 17, with the full Moon on December 24 (though this is best viewed on December 23). We get a second last quarter on December 30.

The Geminids meteor shower peaks in December

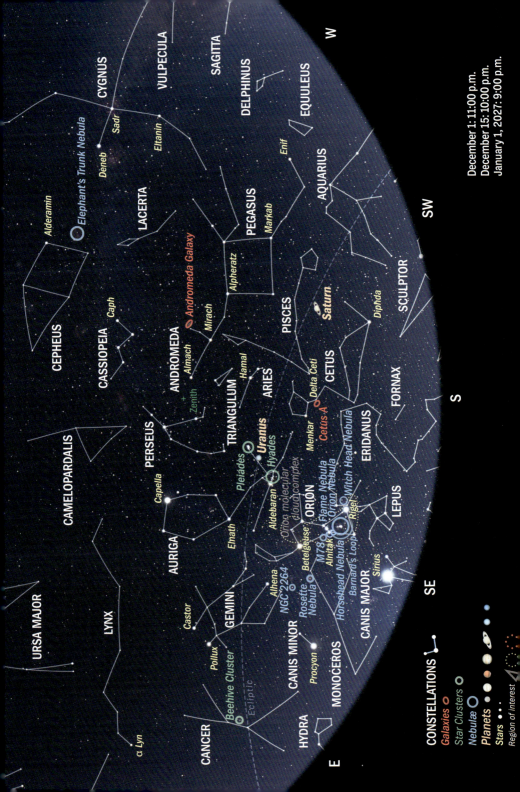

Cetus A (Messier 77)

Highlights in the Southern Sky

Eridanus (the River Eridanus) is now winding its way up the southern sky. Though somewhat difficult to find, it is one of the largest constellations in the night sky, stretching from just east of **Rigel** in **Orion (the Hunter)** to below the southern horizon. In Eridanus, near Rigel, is the **Witch Head Nebula (IC 2118)**, a faint reflection nebula that lies roughly 900 light-years from Earth.

Moving to **Cetus (the Whale)**, you can find **Cetus A (Messier 77)**, a beautiful barred spiral galaxy just near **Delta Ceti**. The galaxy is one of the largest in the Messier Catalogue, with a diameter of roughly 170,000 light-years. It is best viewed through a large telescope.

Looking at Orion, you can find the well-known **Horsehead Nebula (Barnard 33)**. It is part of the **Orion molecular cloud complex**, one of the nearest star-forming regions, lying roughly 1,375 light-years away.

Also nearby is the **Flame Nebula (NGC 2024)**, a bright emission nebula located roughly 1,000 light-years from Earth.

You can find **Messier 78**, a reflection nebula north of the belt star **Alnitak**. Also part of the Orion molecular cloud complex is **Barnard's Loop**, a large emission nebula that looks like a large arc around Orion's belt stars. Though it spans 10 degrees of the sky, it isn't visible, so is best captured in long-exposure photographs.

In **Monoceros (the Unicorn)** is the **Rosette Nebula (NGC 2237)** and its star cluster **NGC 2244**. Also nearby is the **Cone Nebula (NGC 2264)**. It is also home to the **Christmas Tree Cluster**, the **Fox Fur Nebula** and the **Snowflake Cluster**.

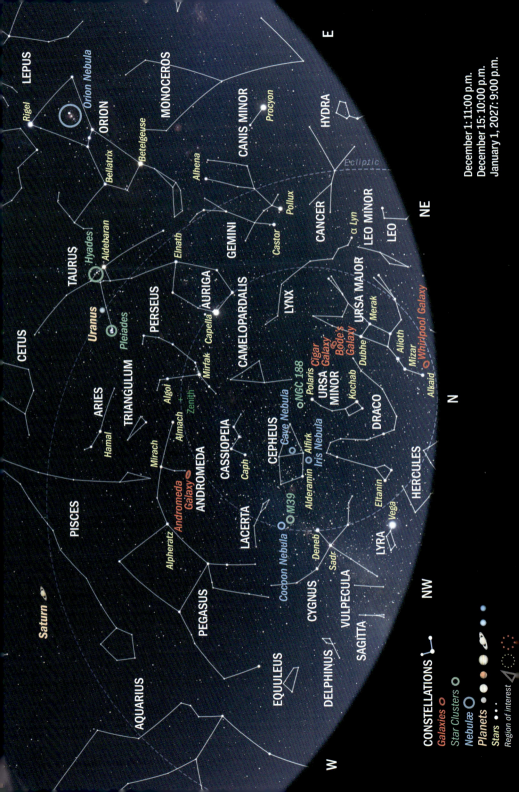

Cave Nebula (Caldwell 9)

December 2026

Highlights in the Northern Sky

In the northwest, **Cygnus (the Swan)** is dipping below the horizon, though its "tail" star **Deneb** is shining brightly. Just north of Deneb is the nice open cluster **Messier 39**. This loose cluster contains at least 30 stars that are estimated to be 200 to 300 million years old.

Nearby is the **Cocoon Nebula (IC 5146)**, a small reflection and emission nebula located roughly 4,000 light-years from Earth. Within the nebula is a cluster of newborn stars, which are responsible for illuminating the nebula's dust and gas.

Moving to **Cepheus (the King)**, you can find the stunning **Iris Nebula (NGC 7023)**. This bright reflection nebula lies 1,300 light-years away and is best viewed through a telescope. It can be found between the stars **Alderamin** and **Alfirk**.

If you have a telescope handy, try to find the **Cave Nebula (Caldwell 9)**, a diffuse emission nebula in Cepheus. The arc that forms the "mouth" of the cave is an active birthplace for stars and of great interest to astronomers who are trying to learn more about how stars begin their lives.

To the northeast is **Ursa Major (the Great Bear)**, where you can find **Bode's Galaxy (Messier 81)** and the **Cigar Galaxy (Messier 82)**. Near the prominent star **Polaris** (though technically still in Cepheus) is the open cluster **NGC 188**. This cluster is one of the oldest-known open clusters, at the ripe old age of 6.8 billion years.

The Eagle Nebula (Messier 16) is a beautiful star-forming region found in the constellation Serpens

The Messier Catalogue

Name	Traditional Name	Type	Constellation
Messier 1 (NGC 1952)	Crab Nebula	Supernova remnant	Taurus
Messier 2 (NGC 7089)		Globular cluster	Aquarius
Messier 3 (NGC 5272)		Globular cluster	Canes Venatici
Messier 4 (NGC 6121)		Globular cluster	Scorpius
Messier 5 (NGC 5904)		Globular cluster	Serpens
Messier 6 (NGC 6405)	Butterfly Cluster	Open cluster	Scorpius
Messier 7 (NGC 6475)	Ptolemy's Cluster	Open cluster	Scorpius
Messier 8 (NGC 6523)	Lagoon Nebula	Emission nebula with cluster	Sagittarius
Messier 9 (NGC 6333)		Globular cluster	Ophiuchus
Messier 10 (NGC 6254)		Globular cluster	Ophiuchus
Messier 11 (NGC 6705)	Wild Duck Cluster	Open cluster	Scutum
Messier 12 (NGC 6218)	Gumball Globular	Globular cluster	Ophiuchus
Messier 13 (NGC 6205)	Hercules	Globular cluster	Hercules
Messier 14 (NGC 6402)		Globular cluster	Ophiuchus
Messier 15 (NGC 7078)	Great Pegasus Cluster	Globular cluster	Pegasus
Messier 16 (NGC 6611)	Eagle Nebula	Emission nebula with cluster	Serpens
Messier 17 (NGC 6618)	Swan Nebula	Emission nebula with cluster	Sagittarius
Messier 18 (NGC 6613)		Open cluster	Sagittarius
Messier 19 (NGC 6273)		Globular cluster	Ophiuchus
Messier 20 (NGC 6514)	Trifid Nebula	Emission, reflection and dark nebula with cluster	Sagittarius
Messier 21 (NGC 6531)		Open cluster	Sagittarius
Messier 22 (NGC 6656)	Great Sagittarius Cluster	Globular cluster	Sagittarius
Messier 23 (NGC 6494)		Open cluster	Sagittarius
Messier 24 (IC 4715)	Sagittarius Star Cloud	Milky Way star cloud	Sagittarius
Messier 25 (IC 4725)		Open cluster	Sagittarius
Messier 26 (NGC 6694)		Open cluster	Scutum
Messier 27 (NGC 6853)	Dumbbell Nebula	Planetary nebula	Vulpecula
Messier 28 (NGC 6626)		Globular cluster	Sagittarius
Messier 29 (NGC 6913)		Open cluster	Cygnus

Name	Traditional Name	Type	Constellation
Messier 30 (NGC 7099)		Globular cluster	Capricornus
Messier 31 (NGC 224)	Andromeda Galaxy	Spiral galaxy	Andromeda
Messier 32 (NGC 221)	Le Gentil	Dwarf elliptical galaxy	Andromeda
Messier 33 (NGC 598)	Triangulum Galaxy	Spiral galaxy	Triangulum
Messier 34 (NGC 1039)		Open cluster	Perseus
Messier 35 (NGC 2168)		Open cluster	Gemini
Messier 36 (NGC 1960)	Pinwheel Cluster	Open cluster	Auriga
Messier 37 (NGC 2099)		Open cluster	Auriga
Messier 38 (NGC 1912)	Starfish Cluster	Open cluster	Auriga
Messier 39 (NGC 7092)		Open cluster	Cygnus
Messier 40	Winnecke 4	Double star	Ursa Major
Messier 41 (NGC 2287)		Open cluster	Canis Major
Messier 42 (NGC 1976)	Orion Nebula	Emission-reflection nebula	Orion
Messier 43 (NGC 1982)	De Mairan's Nebula	Emission-reflection nebula	Orion
Messier 44 (NGC 2632)	Beehive Cluster	Open cluster	Cancer
Messier 45	Pleiades	Open cluster	Taurus
Messier 46 (NGC 2437)		Open cluster	Puppis
Messier 47 (NGC 2422)		Open cluster	Puppis
Messier 48 (NGC 2548)		Open cluster	Hydra
Messier 49 (NGC 4472)		Elliptical galaxy	Virgo
Messier 50 (NGC 2323)	Heart-shaped Cluster	Open cluster	Monoceros
Messier 51 (NGC 5194, NGC 5195)	Whirlpool Galaxy	Spiral galaxy	Canes Venatici
Messier 52 (NGC 7654)		Open cluster	Cassiopeia
Messier 53 (NGC 5024)		Globular cluster	Coma Berenices
Messier 54 (NGC 6715)		Globular cluster	Sagittarius
Messier 55 (NGC 6809)	Summer Rose Star	Globular cluster	Sagittarius
Messier 56 (NGC 6779)		Globular cluster	Lyra
Messier 57 (NGC 6720)	Ring Nebula	Planetary nebula	Lyra
Messier 58 (NGC 4579)		Barred spiral galaxy	Virgo
Messier 59 (NGC 4621)		Elliptical galaxy	Virgo
Messier 60 (NGC 4649)		Elliptical galaxy	Virgo

Name	Traditional Name	Type	Constellation
Messier 61 (NGC 4303)		Spiral galaxy	Virgo
Messier 62 (NGC 6266)		Globular cluster	Ophiuchus
Messier 63 (NGC 5055)	Sunflower Galaxy	Spiral galaxy	Canes Venatici
Messier 64 (NGC 4826)	Black Eye Galaxy	Spiral galaxy	Coma Berenices
Messier 65 (NGC 3623)		Barred spiral galaxy	Leo
Messier 66 (NGC 3627)		Barred spiral galaxy	Leo
Messier 67 (NGC 2682)	King Cobra Cluster	Open cluster	Cancer
Messier 68 (NGC 4590)		Globular cluster	Hydra
Messier 69 (NGC 6637)		Globular cluster	Sagittarius
Messier 70 (NGC 6681)		Globular cluster	Sagittarius
Messier 71 (NGC 6838)		Globular cluster	Sagittarius
Messier 72 (NGC 6981)		Globular cluster	Aquarius
Messier 73 (NGC 6994)		Asterism	Aquarius
Messier 74 (NGC 628)	Phantom Galaxy	Spiral galaxy	Pisces
Messier 75 (NGC 6864)		Globular cluster	Sagittarius
Messier 76 (NGC 650, NGC 651)	Little Dumbbell Nebula	Planetary nebula	Perseus
Messier 77 (NGC 1068)	Cetus A	Spiral galaxy	Cetus
Messier 78 (NGC 2068)		Reflection nebula	Orion
Messier 79 (NGC 1904)		Globular cluster	Lepus
Messier 80 (NGC 6093)		Globular cluster	Scorpius
Messier 81 (NGC 3031)	Bode's Galaxy	Spiral galaxy	Ursa Major
Messier 82 (NGC 3034)	Cigar Galaxy	Starburst irregular galaxy	Ursa Major
Messier 83 (NGC 5236)	Southern Pinwheel Galaxy	Barred spiral galaxy	Hydra
Messier 84 (NGC 4374)		Lenticular or elliptical galaxy	Virgo
Messier 85 (NGC 4382)		Lenticular or elliptical galaxy	Coma Berenices
Messier 86 (NGC 4406)		Lenticular or elliptical galaxy	Virgo
Messier 87 (NGC 4486)	Virgo A	Elliptical galaxy	Virgo
Messier 88 (NGC 4501)		Spiral galaxy	Coma Berenices
Messier 89 (NGC 4552)		Elliptical galaxy	Virgo
Messier 90 (NGC 4569)		Spiral galaxy	Virgo

Name	Traditional Name	Type	Constellation
Messier 91 (NGC 4548)		Barred spiral galaxy	Coma Berenices
Messier 92 (NGC 6341)		Globular cluster	Hercules
Messier 93 (NGC 2447)		Open cluster	Puppis
Messier 94 (NGC 4736)	Cat's Eye Galaxy	Spiral galaxy	Canes Venatici
Messier 95 (NGC 3351)		Barred spiral galaxy	Leo
Messier 96 (NGC 3368)		Spiral galaxy	Leo
Messier 97 (NGC 3587)	Owl Nebula	Planetary nebula	Ursa Major
Messier 98 (NGC 4192)		Spiral galaxy	Coma Berenices
Messier 99 (NGC 4254)	Coma Pinwheel	Spiral galaxy	Coma Berenices
Messier 100 (NGC 4321)		Spiral galaxy	Coma Berenices
Messier 101 (NGC 5457)	Pinwheel Galaxy	Spiral galaxy	Ursa Major
Messier 102 (NGC 5866)	Spindle Galaxy	Lenticular galaxy	Draco
Messier 103 (NGC 581)		Open cluster	Cassiopeia
Messier 104 (NGC 4594)	Sombrero Galaxy	Spiral galaxy	Virgo
Messier 105 (NGC 3379)		Elliptical galaxy	Leo
Messier 106 (NGC 4258)		Spiral galaxy	Canes Venatici
Messier 107 (NGC 6171)		Globular cluster	Ophiuchus
Messier 108 (NGC 3556)	Surfboard Galaxy	Barred spiral galaxy	Ursa Major
Messier 109 (NGC 3992)		Barred spiral galaxy	Ursa Major
Messier 110 (NGC 205)	Edward Young Star	Dwarf elliptical galaxy	Andromeda

The galaxies Messier 95 (bottom) and Messier 96 (top) can both be found in Leo

Glossary

annular solar eclipse: A kind of partial solar eclipse during which the Moon does not completely cover the Sun's disk but leaves a ring of sunlight around the Moon.

aphelion: For an object orbiting the Sun, the point in its orbit that is farthest from the Sun.

apogee: When the Moon (or any satellite of Earth) is at its most distant in its monthly orbit around Earth.

arcsecond (or second of arc): A tiny angle equal to 1/3600 of a degree, used to measure the separation of double stars and the apparent diameters of Solar System objects.

asterism: A group of stars within a constellation (or sometimes from more than one constellation) that forms its own distinct pattern.

astronomical unit (au): A unit of distance that uses the average distance between Earth and the Sun as its base metric. One astronomical unit is approximately 150 million kilometers (93.2 million miles).

autumnal (fall) equinox: The date in September marking the end of summer and the beginning of autumn, when the Sun illuminates the Northern and Southern Hemispheres equally. The lengths of the day and night are equal, and it's one of two instances of the year that the Sun has a declination of 0 degrees.

binary star system: A double-star system in which two stars are gravitationally bound to each other and orbit a common center of mass. Many double-star systems have multiple components.

conjunction: Technically speaking a conjunction is when two objects have the same right ascension, but amateur astronomers also use the term when there is a close approach of two or more celestial objects.

constellation: A group of stars that make an imaginary image in the night sky. The International Astronomical Union (IAU) has designated the boundaries between 88 official constellations covering the celestial sphere.

degree: A unit of measurement used to measure the distance between objects or the position of objects in astronomy. The entire sky spans 360 degrees, and up to about 180 degrees of sky is visible from any given point on Earth with an unobstructed horizon.

double star (binary star): A pair of stars with small angular separation (from fractions of an arcsecond to hundreds of arcseconds). Optical doubles are chance alignments of stars with no physical connection to each other. True binary stars are gravitationally bound (see binary star system).

fireball: A very bright meteor, generally brighter than magnitude −4.0 (about the brightness of Venus).

galaxy: An enormous system of gas, dust and billions of stars and their planetary systems, all held together by gravity.

globular cluster: Old star systems at the edges of spiral galaxies that can contain anywhere from thousands to millions of stars, packed in a close, roughly spherical form and held together by gravity.

greatest eastern elongation: When an inner planet (Mercury or Venus) is farthest from the Sun in the evening sky.

greatest western elongation: When an inner planet (Mercury or Venus) is farthest from the Sun in the morning sky.

light-year: A unit of distance that uses how far light travels in one Earth year as its base metric. One light-year is about 9 trillion kilometers (6 trillion miles). Objects in outer space are so far apart that it takes a long time for their light to reach Earth, and so the farther an object is, the farther in the past that observers on Earth are seeing it. As an example, the Andromeda Galaxy (Messier 31) is 2.5 million light-years away, so when observers look at it in the sky, they are seeing it as it appeared 2.5 million years ago.

magnitude: The apparent brightness of an object in the sky. Magnitude is measured on a scale where the higher the number, the fainter the object appears. Objects with negative numbers are brighter than those with positive numbers.

meteor shower: An atmospheric event during which a number of meteors can be seen radiating from one point in the sky. Meteor showers occur when Earth, at a particular point in its orbit, crosses a stream of particles left over from a passing comet or asteroid.

meteor shower peak: The best time to view a meteor shower, when meteor activity will be at its highest.

nebula: A cloud of dust and gas visible in the night sky either as a bright patch or a dark shadow against other luminous matter. To learn more about the different types of nebulae, see page 42.

occultation: When one object appears to pass in front of another more distant object, obscuring the observer's view. For example, a lunar occultation is when the Moon appears to pass in front of a more distant object, like a planet or star.

open cluster: A star system that contains anything from a dozen to hundreds of stars, all formed from the same hydrogen cloud, in which the stars are spread out. Open clusters are mostly found near the galactic plane, the plane on which most of a galaxy's mass lies.

opposition: When an outer planet (Mars, Jupiter, Saturn, Uranus or Neptune), a dwarf planet or a minor planet is opposite the Sun in the sky in right ascension.

partial lunar eclipse: A lunar eclipse during which the Earth's shadow partially covers the Moon.

partial solar eclipse: A solar eclipse during which the Moon only partially covers the Sun.

perigee: When the Moon (or any satellite of Earth) is at its closest in its monthly orbit around Earth.

perihelion: For an object orbiting the Sun, the point in its orbit that is nearest to the Sun.

star: A luminous ball of plasma held together by its own gravity and fueled initially by the nuclear fusion of hydrogen into helium at its core. Stars come in a vast variety of sizes, luminosities and temperatures, typically ranging from dwarf sizes (that are as little as 10 percent the mass of the Sun) to hypergiants (that can be 100 or more times the mass of the Sun). To learn more about the different types of stars, see pages 14–15.

summer solstice: The point during the year that a particular hemisphere is most tilted toward the Sun. This takes place in June in the Northern Hemisphere.

supernova: A powerful and luminous explosion that occurs during the last evolutionary stages of a massive star when its brightness increases immensely and it throws off most of its mass; a supernova can also be when a white dwarf is triggered into runaway nuclear fusion by the accretion of material from a binary companion; or by a stellar merger.

total lunar eclipse: A lunar eclipse during which the Earth's shadow completely covers the Moon. Such an eclipse can last for hours.

total solar eclipse: A solar eclipse during which the Moon covers the entire face of the Sun, revealing the Sun's corona and prominences. Solar eclipses can last from a few seconds to about seven minutes maximum at a specific location.

variable star: A star whose apparent magnitude varies over time. The variability might be caused by physical changes to the star itself or by how it rotates or interacts with nearby objects.

vernal (spring) equinox: The date in March marking the end of winter and the beginning of spring, when the Sun illuminates the Northern and Southern Hemispheres equally. The lengths of the day and night are equal, and it's one of two instances of the year that the Sun has a declination of 0 degrees

winter solstice: The point during the year that a particular hemisphere is most tilted away from the Sun. This takes place in December in the Northern Hemisphere.

Zenithal Hourly Rate (ZHR): The rate of meteors a shower would produce per hour under clear, dark skies and with the radiant at the zenith.

The Greek Alphabet (lowercase)

α	alpha	ζ	zeta	λ	lambda	π	pi	φ	phi
β	beta	η	eta	μ	mu	ρ	rho	χ	chi
γ	gamma	θ	theta	ν	nu	σ	sigma	ψ	psi
δ	delta	ι	iota	ξ	xi	τ	tau	ω	omega
ε	epsilon	κ	kappa	ο	omicron	υ	upsilon		

Resources

American Astronomical Society — aas.org — An organization of professional astronomers seeking to enhance and share humanity's understanding of the Universe

American Meteor Society — amsmeteors.org — A non-profit scientific organization that informs, encourages and supports research in Meteor Astronomy

Eclipsewise.com — eclipsewise.com — Astronomer Fred Espenak's site dedicated to predictions and information on eclipses of the Sun and Moon

European Southern Observatory — eso.org — An intergovernmental research organization for ground-based astronomy that shares recent research, news and images

European Space Agency (ESA) — esa.int — The main site for the European Space Agency that shares news, research and launches

Hubblesite.org — hubblesite.org — NASA's main site for the Hubble Space Telescope, including news and images

James Webb Space Telescope — webb.nasa.gov — NASA's main site for the James Webb Space Telescope, including the latest images

NASA — nasa.gov — An update on all NASA missions, including astronomical events, news and launches

The Planetary Society — planetary.org — An international, non-profit organization that promotes the exploration of space through education, advocacy and research

The Royal Astronomical Society of Canada — rasc.ca — Home to Canada's national astronomical association

Spaceweather.com — spaceweather.com — An update on space weather, including solar flares, coronal mass ejections and noctilucent cloud forecasts

Space Weather Prediction Center — swpc.noaa.gov — The National Oceanic and Atmospheric Administration's site on forecasts for space weather, including potential geomagnetic storms

Solar and Heliospheric Observatory — soho.nascom.nasa.gov — A collaborative project between the ESA and NASA to study the Sun and its interactions with Earth from different perspectives

Solar Dynamics Observatory — sdo.gsfc.nasa.gov — Satellite observations of the Sun

Photo Credits

Alan Dyer: 31, 71, 97
Alan Dyer / Stocktrek Images / Alamy Stock Photo: 7, 109
Alan Dyer / VW Pics / Alamy Stock Photo: 10, 61
Andrea Girones: 59, 119
Andre Paquette: 63
Artem Povarov / Alamy Stock Photo: 103
David Hoskin: 79
Debra Ceravolo: 5, 18, 43 (top), 101
Elliot Severn: 11, 12, 13
Fred Espenak: 33-35
James Cleland: 55
Katelyn Beecroft: 53, 57
Kimberly Sibbald: 105
Malcolm Park: 21 (bottom), 32, 42, 51, 87, 111, 115
Mark Germani: 69, 81, 89
Michael Watson: 26-27

NASA/Bill Dunford: 24
NASA/ESA/A. Simon (GSFC)/M.H. Wong (University of California, Berkeley)/OPAL Team: 41 (top)
NASA, ESA & A. van der Hoeven: 117
NASA/ESA/CSA/STScI: 9
NASA/ESA/HEIC/The Hubble Heritage Team (STScI/AURA): 77
NASA/ESA/STScI/Hubble: 21 (top)
NASA/ESA/The Hubble Heritage Team (STScI/AURA); Acknowledgment: S. Smartt (The Queen's University of Belfast): 45 (top)
NASA/Goddard/University of Arizona: 19

NASA/Johns Hopkins University Applied Physics Laboratory/Carnegie Institution of Washington: 38
NASA/JPL: 41 (bottom)
NASA/JPL-Caltech: 39
NASA/JPL/Space Science Institute: 40
NASA/SDO: 30 (bottom)
Nicole Mortillaro: 67
Notanee Bourassa: 29, 37
Rob Lyons: 107
Ron Brecher: 65, 75, 83, 95, 99, 113, 120, 125
Sean Walker: 30 (top), 43 (bottom)
Sheila Wiwchar: 85
Shutterstock/Nikta_Nikta: 20
Stefanie Harron: 8, 93
Tim Colwell: 49

"The eyes of the world are upon you. The hopes and prayers of liberty-loving people everywhere march with you."

— Dwight D. Eisenhower

D-Day, the 6th of June, is one of the most significant dates in the calendar. The Twentieth Century's fate hung in the balance, as Nazi Germany had marched across Europe, bringing the blitzkrieg, the Holocaust, and the mad dreams of Adolf Hitler across the bloodied landscape of Europe. France had fallen, and until the Americans joined the battle in December 1941, the balance of power was securely in the hands of the Third Reich. But even as the Allied forces joined together under the leadership of Supreme Allied Commander General Dwight D. Eisenhower, Germany continued to win battles. In order for the Nazis to be stopped, the Allies had to invade occupied Europe. And invade they did, launching the most massive amphibious landing the world had ever seen. The pantheon of heroes of the Normandy invasion includes soldiers who are unknown except for the crosses that bear their names in the cemeteries where they were buried, but all, those who died and those who survived, played their part in the liberation of Europe.